# Landscape Architecture and Design - Sustainability and Management

*Edited by Sérgio Lousada*

Published in London, United Kingdom

Landscape Architecture and Design - Sustainability and Management
http://dx.doi.org/10.5772/intechopen.1004506
Edited by Sérgio Lousada

Assistant to the Editor: José Mário Cabral Fernandes

Contributors
Andrii Khorolskyi, Annissa Muhammed, Ayşenur Begüm Helvaci, Deborah Bunmi Ojo, Denisa Pekna, Dennis Doxtater, Eyasu Elias, Gemma Chiaffarelli, Ilda Vagge, Jana Moravcova, José Mário Cabral Fernandes, Luigino Pirola, Maddalena Gioia Gibellii, Myroslava Lendel, Nicolò Sgalippa, Oksana Sviezhentseva, Raquel Carvalho, Svitlana Delehan, Sérgio Lousada, Vendula Moravcova, Zeynep Yazicioğlu Halu



First published in London, United Kingdom, 2025 by IntechOpen
IntechOpen is the global imprint of INTECHOPEN LIMITED, registered in England and Wales, registration number: 11086078, 167-169 Great Portland Street, London, W1W 5PF, United Kingdom

For EU product safety concerns: IN TECH d.o.o., Prolaz Marije Krucifikse Kozulić 3, 51000 Rijeka, Croatia, info@intechopen.com or visit our website at intechopen.com.

British Library Cataloguing-in-Publication Data
A catalogue record for this book is available from the British Library

Landscape Architecture and Design - Sustainability and Management
Edited by Sérgio Lousada
p. cm.

This title is part of the Sustainable Development Book Series, Volume 31
Topic: Urban Planning and Environmental Management
Series Editor: Usha Iyer-Raniga
Topic Editors: Rui Alexandre Castanho and Luís Loures

Print ISBN 978-0-85466-629-4
Online ISBN 978-0-85466-628-7
eBook (PDF) ISBN 978-0-85466-630-0
ISSN 2753-6580

If disposing of this product, please recycle the paper responsibly.

IntechOpen Book Series
Sustainable Development
Volume 31

## Aims and Scope of the Series

Transforming our World: the 2030 Agenda for Sustainable Development endorsed by United Nations and 193 Member States, came into effect on Jan 1, 2016, to guide decision making and actions to the year 2030 and beyond. Central to this Agenda are 17 Goals, 169 associated targets and over 230 indicators that are reviewed annually. The vision envisaged in the implementation of the SDGs is centered on the five Ps: People, Planet, Prosperity, Peace and Partnership. This call for renewed focused efforts ensure we have a safe and healthy planet for current and future generations.

This Series focuses on covering research and applied research involving the five Ps through the following topics:

1. Sustainable Economy and Fair Society that relates to SDG 1 on No Poverty, SDG 2 on Zero Hunger, SDG 8 on Decent Work and Economic Growth, SDG 10 on Reduced Inequalities, SDG 12 on Responsible Consumption and Production, and SDG 17 Partnership for the Goals

2. Health and Wellbeing focusing on SDG 3 on Good Health and Wellbeing and SDG 6 on Clean Water and Sanitation

3. Inclusivity and Social Equality involving SDG 4 on Quality Education, SDG 5 on Gender Equality, and SDG 16 on Peace, Justice and Strong Institutions

4. Climate Change and Environmental Sustainability comprising SDG 13 on Climate Action, SDG 14 on Life Below Water, and SDG 15 on Life on Land

5. Urban Planning and Environmental Management embracing SDG 7 on Affordable Clean Energy, SDG 9 on Industry, Innovation and Infrastructure, and SDG 11 on Sustainable Cities and Communities.

The series also seeks to support the use of cross cutting SDGs, as many of the goals listed above, targets and indicators are all interconnected to impact our lives and the decisions we make on a daily basis, making them impossible to tie to a single topic.

# Meet the Series Editor

Usha Iyer-Raniga is a professor in the School of Property and Construction Management at RMIT University. Usha co-leads the One Planet Network's Sustainable Buildings and Construction Programme (SBC), a United Nations 10 Year Framework of Programmes on Sustainable Consumption and Production (UN 10FYP SCP) aligned with Sustainable Development Goal 12. The work also directly impacts SDG 11 on Sustainable Cities and Communities. She completed her undergraduate degree as an architect before obtaining her Masters degree from Canada and her Doctorate in Australia. Usha has been a keynote speaker as well as an invited speaker at national and international conferences, seminars and workshops. Her teaching experience includes teaching in Asian countries. She has advised Austrade, APEC, national, state and local governments. She serves as a reviewer and a member of the scientific committee for national and international refereed journals and refereed conferences. She is on the editorial board for refereed journals and has worked on Special Issues. Usha has served and continues to serve on the Boards of several not-for-profit organisations and she has also served as panel judge for a number of awards including the Premiers Sustainability Award in Victoria and the International Green Gown Awards. Usha has published over 100 publications, including research and consulting reports. Her publications cover a wide range of scientific and technical research publications that include edited books, book chapters, refereed journals, refereed conference papers and reports for local, state and federal government clients. She has also produced podcasts for various organisations and participated in media interviews. She has received state, national and international funding worth over USD $25 million. Usha has been awarded the Quarterly Franklin Membership by London Journals Press (UK). Her biography has been included in the Marquis Who's Who in the World® 2018, 2016 (33rd Edition), along with approximately 55,000 of the most accomplished men and women from around the world, including luminaries as U.N. Secretary-General Ban Ki-moon. In 2017, Usha was awarded the Marquis Who's Who Lifetime Achiever Award.

# Meet the Volume Editor

Sérgio Lousada holds an International Ph.D. in Civil Engineering – Hydraulics from the University of Minho (UMinho) in Portugal and an International Ph.D. in Sustainable Territorial Development from the University of Extremadura (UEx) in Spain. He has published articles and books and participated in events mainly in the areas of hydraulics, urban planning and land management. Furthermore, he collaborates with the Research Group on Environment and Spatial Planning (MAOT) - University of Extremadura (UEx), Badajoz, Spain; VALORIZA - Research Center for the Enhancement of Endogenous Resources, Polytechnic Institute of Portalegre (IPP), Portugal; CITUR-Madeira - Centre for Tourism Research, Development and Innovation, Madeira, Portugal; RISCO - Civil Engineering Department of the University of Aveiro, Aveiro, Portugal; OSEAN – *Outermost Regions Sustainable Ecosystem for Entrepreneurship and Innovation*, Madeira, Portugal. Moreover, he holds an International Interuniversity Master's Degree in Ports and Coasts Engineering from the University of Las Palmas de Gran Canaria (ULPGC) and the University of La Laguna (ULL), both in Spain.

# Contents

# Preface

Landscape defines the spaces in which we live, interact, and evolve, shaping not only our environments but also our values, aspirations, and connections to nature. To ensure resilient, sustainable, and meaningful ecosystems, landscape architecture plays an increasingly important role as urbanization increases, climate change quickens and ecological concerns become more complicated.

This book, "*Landscape Architecture and Design – Sustainability and Management*", is a team effort that combines a variety of knowledge, creative viewpoints, and forward-thinking research to rethink how humans engage with the built and natural environments. The book provides a thorough examination of modern landscape architecture through its well-chosen chapters, placing a strong emphasis on environmental stewardship, social justice, and ecological integrity.

The understanding that landscape architecture is a dynamic activity grounded in interdisciplinary knowledge—from ecology and urban planning to art, technology, and social sciences—rather than just a design discipline lies at the core of this work. The use of these methods by landscape architects to improve well-being, create stronger bonds between people and their surroundings, and reshape spaces is highlighted in the book.

Each chapter offers insight into important issues and innovative solutions, from green infrastructure integration and climate-resilient architecture to adaptive reuse of cultural landscapes and sustainable site planning. The examples and case studies offered provide motivation and useful advice for students, academics, practitioners, and policymakers alike.

We invite readers to engage with these pages not only as a source of academic knowl edge but as a call to action—to shape landscapes that honor both people and the planet. May this book foster a shared commitment to creativity, sustainability, and excellence in managing our natural and built environments.

This edited book consists of nine chapters focusing on landscape architecture and design. A brief summary of each chapter is given below.

Chapter 1, "Introductory Chapter: Landscape Architecture and Design – Sustainability and Management", explores how landscape architecture integrates art, science, and sustainability to address climate change, urban growth, and ecosystem degradation through holistic planning, resource management, and ecological restoration.

Chapter 2, "Built Environment as an Extension of Nature: Buildings Adapting to Natural Surroundings and Conditions in National Parks", examines the relationship between architecture and wildlife in national parks, identifying gaps in the renowned buildings' adherence to conservation principles and suggesting degrading criteria based on UN projects and management plans to direct sustainable design.

Chapter 3, "Landscape Dynamics and Biodiversity Conservation in the Ethiopian Highland: Assessing Impacts and Strategies for Sustainability", looks into the significant changes in land cover, biodiversity loss, and soil degradation revealed by this research of Bale Mountains National Park (1985–2017), which calls for conservation, restoration, sustainable agriculture, and community-driven policies to preserve ecological integrity.

Chapter 4, "Landscape Ecology and Ecosystem Services as Landscape Analysis and Assessment Tools for Ecological Landscape Planning", in order to detect ecological weaknesses and provide design solutions that improve sustainability, resilience, and landscape balance, this research of a peri-urban area close to Milan uses landscape ecology and ecosystem services.

Chapter 5, "Mesoamerican Pyramidal Shaped Landscapes as Maps in Urban Ceremonial Sites: "Close Mapping" Analysis of Spatial Homologs", suggests deliberate Olmec and Mayan ceremonial landscape architecture that united regions through shared ritual and spatial orientation by revealing large-scale geometric alignments among Mesoamerican sites using proprietary software and probability assessments.

Chapter 6, "The Role of Elements of Small Folk Architecture in the Czech Landscape", examines the Baroque-era cultural environment of the Czech Republic, highlights minor folk architecture monuments that remain today and evaluates their aesthetic worth in particular microregions despite more extensive landscape changes.

Chapter 7, "The Role of Municipal Ecological Structure in Landscape Preservation", explores how Portugal's municipal ecological structures incorporate green infrastructure into urban development in order to preserve landscapes and natural heritage while fostering resilient, sustainable cities aligned with the SDGs and climate change adaptation.

Chapter 8, "Urban Cultural Heritage under Threat: Strategies for Sustainable Use and Modernisation in Non-EU Contexts", is based on a case study in Ukraine that produced significant energy savings and $CO_2$ reductions through focused interventions. This work offers a sustainable retrofit methodology for heritage buildings that improves energy efficiency without sacrificing historical value.

Chapter 9, "Urban Design and Planning: Integrating Nature and the Built Environment", in addition to addressing global issues like urbanization and climate change, examines urban design and planning in landscape architecture and presents case studies that illustrate creative approaches to resilient, egalitarian, and sustainable urban development.

The chapters in this edited volume examine landscape architecture from a management and sustainability perspective. We hope anyone interested in creating considerate, inclusive, and ecologically conscious landscapes will find these contributions useful. This book seeks to serve as a compass, pointing us toward a balance between human endeavours and the natural systems that support us as we traverse an increasingly uncertain future.

**Sérgio Lousada and José Mário Cabral Fernandes**
UMa - Faculty of Exact Sciences and Engineering,
Madeira University Campus Penteada,
Funchal, Portugal

Chapter 1

# Introductory Chapter: Landscape Architecture and Design – Sustainability and Management

*Sérgio Lousada and José Mário Cabral Fernandes*

## 1. Introduction

Landscape architecture is a field that intricately weaves together art, science, and technology to create sustainable and esthetically pleasing environments. As global environmental issues such as climate change, urban expansion, and ecosystem degradation become more pronounced, landscape architects are increasingly called upon to design spaces that not only serve human needs but also support ecological balance and sustainability [1]. This book, *Landscape Architecture and Design - Sustainability and Management*, serves as a critical exploration of how contemporary landscape architecture addresses these pressing challenges [2].

Sustainability in landscape architecture involves more than just "green design." It encompasses a holistic approach to land-use planning, resource management, and ecological restoration [3, 4]. By integrating ecological principles into landscape architecture, professionals are better able to create resilient spaces that function as both human habitats and ecosystems [5]. This introduction delves into the core themes of sustainability and management within landscape architecture, offering insights into the evolving role of the landscape architect and the tools and methodologies driving this shift.

## 2. Evolving practices in landscape architecture

Over the past few decades, landscape architecture has evolved from a primarily esthetic practice into a multifaceted discipline focused on environmental sustainability, social equity, and economic viability [6]. The challenges posed by rapid urbanization, environmental degradation, and resource depletion have pushed the field to innovate in land-use planning, water management, and ecological restoration [7].

Incorporating geoinformatics has revolutionized the way landscape architects analyze spatial data and make informed decisions [8]. These technologies, coupled with advanced modeling tools, allow for the precise planning of landscapes that are both functional and sustainable. Furthermore, the work of Lousada [9] on land-use management highlights the importance of integrating sustainability into long-term urban planning strategies. Cities and rural areas alike are facing the dual pressures of population growth and environmental conservation, necessitating approaches that can balance human needs with ecological health.

## 3. Sustainability in landscape architecture

Sustainability is the cornerstone of contemporary landscape architecture. It is rooted in the principle that the needs of the present should be met without compromising the ability of future generations to meet their own needs [10]. Landscape architects play a vital role in this paradigm by designing spaces that are environmentally sound, economically feasible, and socially inclusive.

One of the most critical challenges in sustainable landscape architecture is water management. Climate change is exacerbating issues such as droughts, flooding, and water scarcity, particularly in urban areas. Sustainable urban drainage systems (SUDS), rain gardens, and green roofs are becoming integral components of modern urban landscapes, providing both functional and esthetic benefits while mitigating the effects of climate change [11]. In Madeira Island, for instance, hydraulic planning has been crucial in adapting to the unique challenges posed by insular territories, as demonstrated by Lousada et al. [12].

The role of vegetation in sustainable landscapes cannot be overstated. As Lousada and Gómez [13] have shown in their analysis of vegetation dynamics in Galicia, Spain, maintaining biodiversity and restoring degraded ecosystems are essential to the health of both natural and human-made environments. Vegetation not only provides esthetic and recreational value but also plays a critical role in carbon sequestration, soil stabilization, and the enhancement of air and water quality [14].

In addition to environmental benefits, sustainable landscapes offer social and economic advantages. Green spaces contribute to mental and physical well-being, provide opportunities for recreation and social interaction, and can increase property values. As demonstracted in sustainable regional planning in Portugal [15], integrating sustainability into landscape architecture also contributes to long-term economic viability by reducing maintenance costs and enhancing the resilience of infrastructure.

The growing recognition of the importance of sustainability has significantly shaped the field of landscape architecture and design in recent years. Landscape architects now face the dual challenge of creating functional, esthetically pleasing spaces while also promoting environmental resilience and sustainability. With the ongoing impacts of urbanization, climate change, and population growth, it has become imperative to integrate sustainability principles into both urban and rural landscapes [2].

**Figure 1.**
*Sustainable urban landscape (Source: [16]).*

One of the key aspects of modern landscape design is the use of green infrastructure to enhance urban resilience. **Figure 1** below illustrates a sustainable urban landscape incorporating features such as green roofs, rain gardens, and modern water management systems. These elements not only beautify the city but also mitigate urban heat islands, reduce stormwater runoff, and support biodiversity.

## 4. Managing landscapes for future resilience

The future of landscape architecture lies in its ability to adapt to new environmental challenges. As the global population continues to rise and urban areas expand, the demand for sustainable, well-managed landscapes will only increase. Landscape architects are uniquely positioned to address issues such as climate resilience, urban heat islands, and the loss of biodiversity through innovative design solutions [4].

A growing area of focus is the integration of technology in landscape management. Tools such as Geographic Information Systems (GIS), remote sensing, and environmental modeling are enabling architects to better understand the long-term impacts of their designs. These technologies provide data that can inform decision-making processes, helping to create landscapes that are not only beautiful but also sustainable and resilient in the face of future challenges [2].

The case of land-use changes in the Alto Tâmega region in Portugal [15] exemplifies the importance of sustainable planning for long-term resilience. By understanding historical land-use patterns and predicting future changes, landscape architects can design spaces that are adaptable to environmental fluctuations, ensuring their relevance and functionality for years to come.

Beyond technology, landscape architecture is also embracing new philosophical approaches to land management. Concepts such as regenerative design, biophilic design, and circular economy principles are gaining traction as they offer more sustainable and ecologically sensitive approaches to the built environment. These approaches emphasize the importance of designing with nature, rather than against

**Figure 2.**
*Rural landscape that balances agricultural productivity with ecological preservation (Source: [17]).*

it, and seek to create landscapes that regenerate ecological systems and enhance biodiversity [1]. Rural landscapes also play a crucial role in sustainable development, especially when it comes to land-use management. **Figure 2** shows a rural landscape that balances agricultural productivity with ecological preservation. The scene includes terraced farming, preserved forests, and eco-friendly structures like solar panels and wind turbines, showcasing the integration of renewable energy sources with land-use practices.

## 5. Conclusion

*Landscape Architecture and Design - Sustainability and Management* aims to provide readers with a comprehensive understanding of how the discipline of landscape architecture is evolving to meet the environmental, social, and economic challenges of the twenty-first century. By drawing on recent advances in geoinformatics, land-use management, and hydraulic planning, this volume offers both theoretical insights and practical solutions for creating sustainable landscapes. It is through this lens of sustainability and management that landscape architects can continue to shape environments that are resilient, beautiful, and sustainable for future generations.

## Author details

Sérgio Lousada[1,2,3,4,5*] and José Mário Cabral Fernandes[1,6]

1 Department of Civil Engineering and Geology (DECG), Faculty of Exact Sciences and Engineering (FCEE), University of Madeira (UMa), Funchal, Portugal

2 VALORIZA—Research Centre for Endogenous Resource Valorization, Portalegre, Portugal

3 RISCO—Civil Engineering Department of University of Aveiro, Aveiro, Portugal

4 CITUR—Madeira—Centre for Tourism Research, Development and Innovation, Madeira, Portugal

5 Research Group on Environment and Spatial Planning (MAOT), University of Extremadura, Badajoz, Spain

6 Faculty of Economics, University of Porto (FEP), Porto, Portugal

*Address all correspondence to: slousada@staff.uma.pt

## References

[1] McHarg IL. Design with Nature. Hoboken, NJ, USA: Wiley; 1995

[2] Benedict MA, McMahon ET. Green Infrastructure: Linking Landscapes and Communities. Washington, DC, USA: Island Press; 2012

[3] Thompson IH. Ecology, Community and Delight: Sources of Values in Landscape Architecture. New York, NY, USA: Routledge; 2011

[4] Corner J. Recovering Landscape: Essays in Contemporary Landscape Architecture. New York, NY, USA: Princeton Architectural Press; 1999

[5] Van der Ryn S, Cowan S. Ecological Design. Washington, DC, USA: Island Press; 2007

[6] Loures L, Panagopoulos T. Sustainable land planning: A collaborative approach. Sustainability. 2010;**2**(11):3572-3592

[7] Newman P, Beatley T, Boyer H. Resilient Cities: Responding to Peak Oil and Climate Change. Washington, DC, USA: Island Press; 2017

[8] Lousada SAN. Geoinformatics in Support of Urban Politics and the Development of Civil Engineering. Hershey, PA, USA: IGI Global; 2023

[9] Lousada SAN. Land-Use Management—Recent Advances, New Perspectives, and Applications. London, UK: IntechOpen; 2024

[10] Ahern J. From fail-safe to safe-to-fail: Sustainability and resilience in the new urban world. Landscape and Urban Planning. 2011;**100**(4):341-343

[11] Waldheim C. Landscape as Urbanism: A General Theory. Princeton, NJ, USA: Princeton University Press; 2016

[12] Lousada S, Gómez JMN. Analyzing the evolution of land-use changes related to vegetation, in the Galicia Region, Spain: From 1990 to 2018. In: Vegetation Dynamics, Changing Ecosystems and Human Responsibility. London, UK: IntechOpen; 2022. DOI: 10.5772/intechopen.106015

[13] Lousada S, Cabezas J, Castanho RA, Gómez JMN. Hydraulic planning in insular urban territories: The case of Madeira Island—Ribeira Brava. Water. 2021;**13**:2951. DOI: 10.3390/w13212951

[14] Spirn AW. The Granite Garden: Urban Nature and Human Design. New York, NY, USA: Basic Books; 1984

[15] Lousada S, Gómez JMN, Loures L. The evolution of land-use changes in the Alto Tâmega Region, Portugal: From 1990 to 2018—A vision of sustainable planning. In: Sustainable Regional Planning. London, UK: IntechOpen; 2023. DOI: 10.5772/intechopen.110036

[16] The City of the Future: Sustainable Urban Development. 2024. Available from: https://www.activesustainability.com/construction-and-urban-development/city-future-sustainable-urban-development/

[17] Ministerie van Landbouw, Natuur en Voedselkwaliteit. Nature-Inclusive Farming: Back to Basics—Advantages of Nature-Inclusive Agriculture—Agrospecials [Webpagina]. Ministerie van Landbouw, Natuur en Voedselkwaliteit; 2024. Available from: https://magazines.rijksoverheid.nl/lnv/agrospecials/2023/01/flip-van-koesveld

Chapter 2

# Built Environment as an Extension of Nature: Buildings Adapting to Natural Surroundings and Conditions in National Parks

*Zeynep Yazicioğlu Halu and Ayşenur Begüm Helvaci*

## Abstract

As the popularity of nature-based tourism rises, it is putting more pressure on natural surroundings, especially on national parks. This study examines the interaction between national parks' built environment and nature. Reviewing global studies, we have developed degradation criteria for national parks and upper-scale design decisions that integrate well with the natural environment. These criteria identify the challenges of increasing leisure activity and provide potential solutions to mitigate these challenges and promote a more sustainable future for our natural areas. Our research included investigating the United Nations' (UN) ten flagship restoration projects and analysing national park management plans in protected natural areas. After reviewing the UN's projects and general management plans, criteria are proposed for examining national parks. As a case study, three architectural designs were chosen from ArchDaily. These buildings were located in national parks and were selected as the building of the year. As a result, it is seen that architectural designs, even though they succeeded as buildings in harmony with nature, did not fit degradation principles, and neither had a general management plan.

**Keywords:** adaptation, architectural design criteria, built environment, management plans, national parks, natural areas, recreation and degradation

## 1. Introduction

As people increasingly seek to escape urban environments and reconnect with nature, the popularity of nature-based tourism is on the rise. This tourism type might contribute to sustainable development. However, the multi-sourced risks to nature should be perfectly analysed and controlled [1]. This trend exerts significant pressure on natural surroundings, particularly on national parks, and underscores the urgent need for immediate action to mitigate potential adverse impacts. Butler argues that tourism under the term sustainability is misleading, and in sensitive areas, restrictive precautions should be taken to minimise tourism activities [2]. On the other hand, developing nations might benefit from ecotourism by developing their tourism strategies for socio-economic development [3]. But in both cases, the role of the tourist is vital.

According to the International Union for Conservation of Nature (IUCN), national parks are natural protected areas with national value and recreational and visitor opportunities [4]. Park management should focus on maintaining biodiversity and carefully developing recreational facilities to avoid diminishing visitor experiences. Overdevelopment could harm both ecological and visitor satisfaction [5]. Thus, when designing national parks, it is crucial first to identify the human and recreation impacts on these vulnerable sites and evaluate the degradation risks; secondly, develop a general framework of upper-settlement decisions and planning policies for long-term developmental plans against those risks; and finally, define a set of architectural design criteria for the park to develop within respect and harmony with nature.

The body of research on national parks and park management is extensive. For example, Moore's manual is an early example that attempts to establish a national park management system and addresses some design issues [6]. Agee and Johnson authored a comprehensive book outlining strategies for ecosystem management in parks [7], while Eagles and McCool investigated tourism management methods in national parks [8]. Prato and Fagre attempt to explain approaches for addressing social, economic and ecological uncertainty [9]. Numerous studies have investigated particular global issues or obstacles in various national parks and tested different management strategies. These studies have concentrated on diverse topics within specific park systems.

Some researchers have explored the tourism aspect of park management, examining stakeholder perceptions, challenges related to tourism capacity, tourism development and the quality visitor experience [10–14]. Several studies have examined different aspects of sustainable development principles in park management [15, 16], the effects of changing population dynamics [17, 18], climate change adaptation [19, 20] and the implications of science-based management approaches [21].

These studies have a narrow focus, examining national parks' specific physical or managerial features. For instance, some studies have analysed the management of specific water bodies within national parks and the effectiveness of particular environmental management plans. Many of these papers utilise case studies of a single park within a park system to address broader topics, such as the trade-offs between development and biodiversity protection [22], the evolution of a country's park management model [23], the practicality of a specific approach to conservation management [24], the relationship between visitor motivation and attitudes towards management restrictions [25] and the development of design criteria for urban parks based on sustainability [26].

In a previous study that compared park management, the survey included managers at the park level but left out those managing the central national park system [27]. The study identified significant issues, including threats to national resources, prominent threats from illegal animal life removal and the lack of personnel. However, given the increasing and worsening environmental issues, these resources are now somewhat outdated and incomplete in summarising the challenges. More importantly, many of these studies are based solely on one country or continent's experience with park management. Little to no work has investigated the UN's ten flagship restoration projects and analysed the relation between national park management plans in protected natural areas and proposed criteria for examining national parks with an architectural overview. There is a gap in the research field that discusses parks with their management plans and architectural practice. There are many contributions to how natural park management plans should be developed. Still, the relationship between nature and the built environment is not discussed deeply, according to buildings' adaptation capabilities to natural surroundings and conditions in national parks. Four architectural designs were chosen from ArchDaily, an online architectural website, to fill this gap, as a case study. These buildings, located in national parks, were selected as the Building of the Year by website users. Their relation to nature and their adaptation capabilities are discussed.

Our motivation in this study is to reveal the extent to which natural areas are under threat due to human activity. In addition, we are motivated to help identify threats and opportunities in advance by creating a general framework for management plans of national parks in the design stage. Thus, it was intended to emphasise that natural areas should be addressed holistically, from deciding whether or not to open natural areas to construction to the principles to be applied on an architectural scale.

When it comes to positioning our approach in the literature, the most noticeable work related to our approach is given by the United Nations (UN), which has declared the ten years between 2021 and 2030 as the Decade of Ecosystem Restoration to prevent, stop and reverse destruction worldwide and developed a Restoration Initiative platform, under United Nations Environment Programme (UNEP). Under this platform, countries can share UNEP's projects on restoring destroyed areas and apply to be selected as a leading project. Ten pioneering projects were first announced at the end of 2022 [28]. Within the scope of this study, these ten pioneering projects were examined regarding the destruction of natural areas and the human-recreation relationship.

Based on our contributions by examining ten pioneering UN restoration projects on natural areas, a scale is proposed to determine the degree of destruction of national parks. In addition, by reviewing the general management plans of four countries, this study reveals the upper-scale decisions that need to be determined for the design of national parks. National parks that have undergone construction and have architectural structures claimed to be of high quality have been evaluated under the suggested headings. Thus, it was possible to question the deterioration of these natural national parks in which they are located and the need for a general management plan.

## 2. Method

When the methods used in studies on national parks were examined, it was determined that case studies were the most frequently used method. Some researchers have analysed management systems of national parks by utilising case studies of a single park within a park system to address broader topics [12, 17, 22–24]. Some case studies deal with the subject in terms of visitor motivation and attitudes towards management systems [13, 14, 25], the development of design criteria according to case studies [26], and surveys according to management systems [16, 27] stand out.

Our study is underpinned by a robust theoretical framework, specifically content analysis, a qualitative research method that meticulously dissects the content of written and visual materials. This method was instrumental in gathering comprehensive information about nature and the built environment, ensuring the thoroughness and credibility of our research.

We examined the protection of natural areas and the relationship between recreation and degradation in natural regions according to the UN's ten flagship restoration projects. Afterwards, a literature review examined various countries' management and planning principles and the design guides they created for national parks. It has been determined that the legal process and requirements of protected areas and recreation-based elements effectively increase the possibility of degradation when designing a national park. They are examined under six main headings: fundamental planning principles, national identity, settlement decisions, local communities, project proposals and facilities and infrastructure.

In line with the study's scope, the subject was approached based on existing regulations and precedent structures. The theoretical framework examines natural protected areas in the context of the UN through regulations and national park definitions. By reviewing the ten pioneering restoration projects of the UN, the reasons for the possible adverse effects that may arise from recreation in national parks were examined. In this context, a scale was prepared to evaluate the relationship between degradation and vulnerability in the parks.

To determine the main headings effective in designing national parks, the general management plans of countries from four different continents, namely the USA, South Africa, Albania and New Zealand, and the plans of one of their national parks were examined to see how these principles were implemented. The general management plan for Albania has yet to be reached.

In this context, ten flagship restoration projects that the UN chose in damaged ecosystems were investigated to identify the possible threats related to recreation. Then, four countries' national park management plans from four diverse continents were analysed to define a general framework for settlement and development decisions of national parks. Finally, three precedent buildings were selected from protected natural areas. The Building of the Year in ArchDaily was evaluated to set architectural design criteria for buildings adapting to the natural surroundings and conditions. This perspective was developed with a comprehensive literature research of degradation possibilities, developmental decisions and case studies embracing the relation with nature.

We draw insightful conclusions from the information we gathered through the buildings. By reviewing the buildings, it has been suggested that the design criteria that are effective in the design of facilities designed in such protected areas to adapt to natural surroundings and conditions can be structured under the following headings: *integrating with the topography, maintenance of the landscape, visual balance of the facade design and the size of the project, benefit from alternative systems for infrastructure.* We tried to gather comprehensive information about the selected projects' degradation possibility and general planning process and draw insightful conclusions from this information about the insights of architectural awards voted by the media.

## 3. Framework for built environment as an extension of nature in national parks

Based on the IUCN, national parks are natural, protected areas with significant national value and opportunities for recreation and visiting [4]. Designing these parks is a complicated process that includes identifying the impacts of human activity and recreation, assessing the risks of degradation and establishing an overall framework for making decisions and planning policies. This framework is essential for developing long-term plans to reduce these risks effectively.

### 3.1 UNEP's ten pioneering restoration projects

UN, in cooperation with Food and Agriculture Organisation (FAO), a UN agency, and UN Environment and Programme (UNEP), has declared the ten years between 2021 and 2030 as the Decade of Ecosystem Restoration to prevent, stop and reverse destruction worldwide. This global response to the destruction and loss of habitats will focus on building the political will and capacity to repair people's relationship with nature [29]. Additionally, the 15th Article of the 2030 Sustainable Development Goals aims to protect and restore regional ecosystems, sustainably manage forests, combat desertification and reverse vegetation and land degradation and biodiversity loss [30]. In this regard, many programs are encouraged by organisations within the UN.

A critical step in creating a shared vision of ecosystem restoration is to adopt principles that underpin the complete set of ecosystem restoration activities. To this end, the UN presented ten principles for ecosystem restoration, including a first principle that orients restoration in the context of the UN Decade, followed by nine best-practice principles. These best-practice principles are published as "Principles for Ecosystem Restoration to Guide the United Nations Decade 2021–2030" and detail the essential tenets of ecosystem restoration that should be followed to maximise net gain for native biodiversity, ecosystem health and integrity, and human health and well-being, across all biomes, sectors and regions [31]. The ten principles can be summarised as "Global Contribution, Broad Engagement, Many Types of Activities, Benefits to Nature and People, Addresses Causes of Degradation, Knowledge Integration, Measurable Goals, Local and Land/Sea Scape Contexts, Monitoring and Management and Policy Integration".

The Restoration Initiative platform, where countries can share UNEP's projects on restoring degraded areas and apply to be selected as a leading project, uses these principles when evaluating. Ten pioneering projects were first announced at the end of 2022. Within the scope of this study, these ten pioneering projects were examined in terms of the destruction of natural areas under the principle Addresses Causes of Degradation. In this context, ten flagship restoration projects that the UN chose in damaged ecosystems were investigated to identify the possible threats related to recreation. These restoration projects are the Atlantic Forest Pact, Abu Dhabi Marine Restoration, Green Great Wall, Namami Gange, Multi-Country Mountain Flagship, Small Island Developing States, Altyn Dala Conservation, Central American Dry Corridor, Building with Nature in Indonesia and Shan-Shui in China.

To ensure the balance between protection and use in national parks, the requirements for recreational areas and the relationship between human activities and the resulting degradation should be prioritised in the design of these parks. In this regard, the ten pioneer restoration projects selected by the UN have been examined based on human activities, creating a scale through activities that cause damage in these areas, which is shared in **Figure 1**. It assesses sensitivity rates and the probability of degradation. The more sensitive an area is regarding its resource values, the more it should be protected, and substances that increase the likelihood of degradation should be kept away from this area as much as possible.

The criteria for evaluating the relationship between degradation and vulnerability in national parks are maximum protection levels and probable degradation, as shown in **Figure 1**. Maximum protection is achieved when *the Values of Natural Areas (Biodiversity, Geological and Value) are high and the Discovery Rate* is low. Maximum probable degradation occurs when *Various Uses (Agriculture, Urbanisation, Grazing*

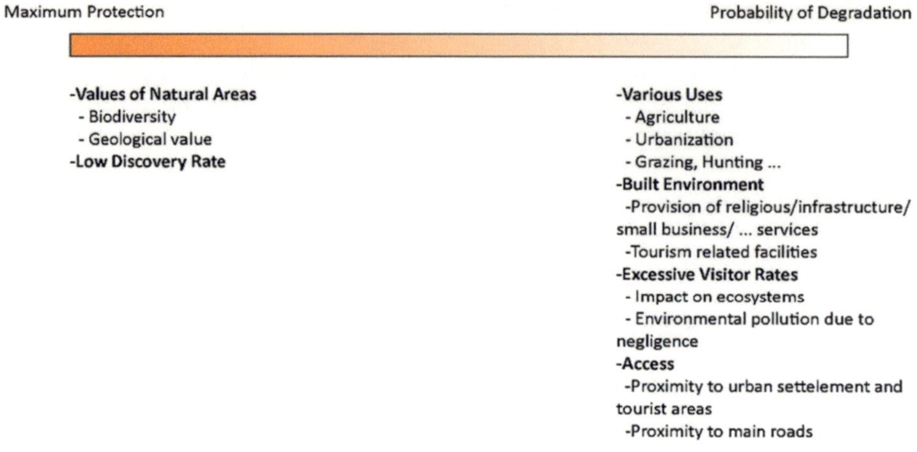


**Figure 1.**
*Probability of degradation scale.*

*and Hunting), Built Environment, Excessive Visitor Rates (Impact on ecosystems, environment pollution) and Access (Proximity to urban settlement and tourist areas, Proximity to main roads)* occur.

Considering national parks as an integrated system, such a ranking could serve as a criterion in the developmental decisions of national parks and the design of specific settlement plans. Various uses such as agriculture, urbanisation, grazing, hunting, etc., construction of new buildings that need infrastructure or new services such as religious buildings, small businesses and tourism-related facilities, excessive visitor rates that have effects on ecosystems and cause pollution by human negligence and easy access to areas by the proximity to urban areas and main roads are essential causes that raise the potential of a natural area to be degraded. Thus, an area with crucial natural values and low discovery rates should maintain these values and avoid these causes as much as possible.

### 3.2 General management plans of four countries

To develop a proposal for the design of structures in national parks, it is necessary first to determine the approach to the parks and developmental decisions in the park's layout scheme. Many assessment and certification systems, such as Leadership in Energy and Environmental Design (LEED), Building Research Establishment's Environmental Assessment Method (BREEAM), Green League and Environmental and Social Responsibility Index, have been widely implemented in national parks over the last decade. These systems aim to assess and certify the sustainability performance levels of national parks by minimising environmental damage, designing sustainable environments and establishing sustainable planning and ecological design criteria. They play an increasingly crucial role in objectively evaluating negative environmental impacts. These evaluation systems define and measure variables related to land use and biodiversity, master plan, building design, resource usage, research and information technologies, carbon targets, sustainability responsibilities, waste management, recycling, local emissions, users, transportation, local integration and social integration. Additionally, these systems allow for comparisons between built environments [32–34].

Developing general management plans for environments is also a key attribute these certification systems use. In this context, general management plans of four countries' national parks and specific management plans of one national park from each country have been examined. Four countries from diverse continents were selected according to accessibility to comprehensive data. From the selected countries, the United States focuses on preserving biodiversity, recreational opportunities, visitor experience and stakeholder participation. New Zealand prioritises indigenous rights, ecosystem restoration in alignment with conservation principles and public involvement. South Africa adopts an integrated approach combining conservation with socio-economic development and employs management strategies to address challenges. Albania is transitioning its park management with new initiatives to balance preservation with tourism development and infrastructure enhancement.

National park management plans of the United States, South Africa, Albania, and New Zealand have been reviewed [35–40]. Subsections outlining design considerations for national park management plans were evaluated and organised into primary categories, which were classified into developmental decisions and presented in this section. These categories are the findings and conclusions drawn from examining the ten pioneering UN restoration projects in natural areas: fundamental planning principles, national identity, settlement decisions, local communities, project proposals and *facilities and infrastructure*. The items shared here do not encompass all content considered when creating a management plan for national parks but instead focus on decisions that are impactful to design.

*Planning principles*: National parks should have clearly defined reasons and objectives for their establishment, guiding their long-term development plans. These plans should outline the purpose, goals, criteria for development scenarios, implementation strategies, limitations, potential threats and action plans to address these threats. Planning should be flexible to accommodate future scenarios and ecosystem changes. A shared vision should be established for the park, fostering partnerships to ensure management aligns with goals. Changes to park plans should prioritise human safety, ecosystem restoration and resource protection. Recreational activities should align with park values, and visitors should be educated about these activities [35–40].

*National identity*: National parks play a crucial role in forming a national identity through their complementary nature and contribution as part of a comprehensive system. Each park's planning should harmonise with and support the activities of other parks. Also, adopting a common design language across parks and using it consistently in materials, promotions and signage enhances their overall value [35–40].

*Settlement decisions*: Zoning decisions for national parks must balance nature conservation, sensitivity levels and activity interactions. Sensitivity analyses are crucial for identifying conservation values, development zones, included activities and conservation goals. Land use plans define areas needing protection, zones unsafe for human intervention, wildlife boundaries, buffer zones, intervention areas, nature experience zones, visitor areas, facility locations, entry points and transportation routes. Unused park structures, like visitor centres, should be repurposed; unnecessary ones should be removed to reduce operating costs. Non-essential activities should benefit local communities outside the park, minimising development zones. Visitor centres are vital for interpretive services but should avoid unacceptable impacts inside the park. Energy, water and waste management systems should ideally be located outside to minimise park impact [35–40].

*Local people*: Supporting the local community's economy within and near national parks promotes regional development and creates job opportunities. This fosters a sense of ownership and belonging among locals, increasing conservation awareness and contributing to their livelihoods. Local communities should actively participate in all park planning and operation stages, contributing cultural knowledge and aiding conservation efforts. Certain activities beneficial to local communities, like grazing, logging and hunting, should be permitted in parks with careful management to avoid overuse and ensure sustainability [35–40].

*Project proposals*: Before approving a new project in a national park, it should be demonstrated that services cannot be provided feasibly outside park boundaries, and non-construction alternatives should be considered first. If the project proceeds, it must meet several conditions: alignment with the park's purpose and goals, non-interference with existing activities, compatibility with the park's concept, enhancement of visitor experience and consideration of environmental sensitivity in design. It should avoid accelerating natural erosion or negatively impacting waterways, minimise adverse effects, restore damaged vegetation and adhere to sustainable landscape design practices. The project should limit its footprint, complement existing facilities, maintain naturalness, ensure financial sustainability and enhance accessibility, educational opportunities and scenic values. It should also introduce new recreational opportunities, improve circulation and enhance the visibility and security of park values. Land selection for development should align with the park's activities and goals as outlined in the plan. Any development outside this plan should involve removing constructed structures, rehabilitating damaged areas, restoring natural topography, replanting vegetation and restoring natural processes and conditions [35–40].

*Facilities and infrastructure*: Facilities within national parks should be provided to cater to visitors' needs, management and species conservation while aligning with the park's design. During planning and building design, public participation should be supported, life cycle costs evaluated, integration with the park's natural environment ensured, adverse impacts minimised and efficient transportation systems utilised. Infrastructure services for facilities should be minimised, and transportation routes should be strategically located to reduce environmental impact and maintain natural tranquillity. The type of roads used (asphalt, gravel, etc.) should be chosen based on the specific activity served, usage rates and environmental considerations [35–40].

## 4. Case studies: Buildings in harmony with nature

To prevent environmental degradation in national parks, facilities must be determined in park management plans and protective measures implemented. This involves analysing the potential for degradation and sensitivity levels and establishing architectural design criteria for buildings within these areas. In this context, three projects awarded by an online architectural platform, ArchDaily, as the Best of the Year (BOTY) were analysed according to their degradation and protection rates under *natural area values, discovery rate, various land uses, built environment, visitor rates and access roads*. The planning process and its results were researched according to *fundamental planning principles, national identity, settlement decisions, local communities, project proposals and facilities and infrastructure*. Finally, global precedents were evaluated according to their properties, adapting to natural surroundings and conditions.

ArchDaily (www.archdaily.com), a website established in 2008 to showcase the latest architectural developments, was used as a source. Three buildings located in national parks were selected from projects voted BOTY by users annually. This selection method was chosen due to ArchDaily's extensive and accessible content worldwide.

These three projects were analysed according to their approach to nature, and as a result, criteria that affect buildings' relationship with the natural environment were identified.

### 4.1 Fogo Natural Park venue by OTO, 2013

Fogo Natural Park is formed by an active volcano, its crater and rims, and a forest. There is only one settlement in the middle of the caldera, Chã das Caldeiras, which makes a living mainly from farming activities and tourism. The region is one of the poorest in the country, with undeveloped roads and infrastructure systems and a lack of public facilities [41].

The park stands out with its biodiversity and geological properties. It was declared an important bird area in 2022 [42] and nominated as a biosphere reserve in 2020 [43]. Also, being an active volcano, its unique landscape holds significant tourist geological hotspots. Under the United Nations Development Programme, there are regional projects raising awareness and educating locals about biodiversity and protecting ecosystems [44], which might increase the discovery rate of the park. Also, being featured on the most visited architectural platform and winning an award for the visitor centre in the park raises recognition. In Krakow University, an architectural student project was studied on the location of the destroyed building, which indicates that the region is getting recognition [45].

The natural area's values suggest a high level of protection, and the fact that the region needs to be well-developed means it has the potential to sustain that level. However, illegal activities in the area, such as agricultural land use and overgrazing, put the park in a dangerous position, making it open to degradation.

The Fogo Natural Park Venue project (**Table 1**) brings together the locals and park management to solve these issues by creating a common ground for them and offering spaces for both management and recreation. Although the planning of the project could not see the future threats and got destroyed by the volcano [47], its efforts to benefit people were significant in planning, and the fact that the building was not dependent on active energy systems and infrastructure proves that. However, not foreseeing the potential threats and wasting crucial resources in constructing the building where the location was unsafe show a lack of planning.

The region has one of the lowest incomes in the country, so local people use Fogo National Park's resources illegally. To solve these conflicts between locals and park management, the project was designed as two blocks: recreational and management. In the recreational part, there are spaces for both visitors and locals [46].

A prominent feature of this building is its adaptive relationship with the natural topography. Conforming to the existing land contours minimises the need for extensive excavation and grading, thereby preserving the natural landscape and reducing soil erosion. This approach enhances the aesthetic appeal and ensures a symbiotic relationship with the environment. Another essential feature of this building is its green roofs, which continue the existing vegetation. This way, the building contributes to biodiversity, improving air quality and aesthetics by adapting to natural surroundings.

Fogo Natural Park Venue also integrates with the landscape visually by consisting of concrete blocks that use the ashes from the nearby volcano, which gives

| Fogo Natural Park venue | Visitor centre | |
|---|---|---|
| Architect: OTO<br>Total area: 3200 $m^2$<br>Year: 2013<br>Location: Fogo, Cape Verde |  | |
| Architectural Program:<br>The building is divided into two main programs. The cultural area includes an auditorium, open-air theatre, library and terrace café. The administration section has meeting rooms, offices, laboratories and technical areas. | Interaction with Nature:<br>The architecture and natural environment were intended to complement each other.<br>Ashes from the volcano were used in the mixture of wall blocks, ensuring visual unity with the environment.<br>Bright light was avoided to protect the birds in the area. | Design and Spatial Set-Up:<br>The building mass is designed as a continuous shell. Shadow plays were provided with the long walls in the building, ensuring the integration of the building with the road.<br>The building was designed to be self-sufficient since the land had no infrastructure. The inclined surfaces of the building continue from the roof to the central courtyards, creating open spaces. The plant species that define the national park are planted on these surfaces, ensuring integration with the environment. |
| 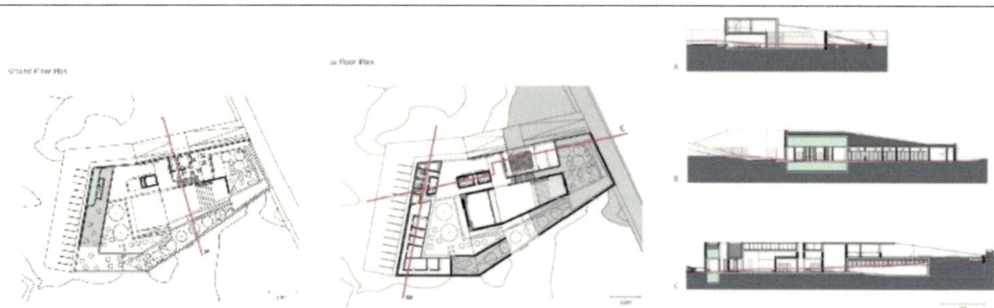 | | |
| Green Building Concept:<br>Solar panels on the roof provide the building's energy needs, and natural ventilation is used. Grill systems integrated into the facade transfer heat during the day, and natural ventilation is provided at night. The building collects and uses rainwater and greywater technology. | Building and Construction System:<br>The building is built by the local people, for the local people, using local materials and techniques.<br> | Facade and Shell Design:<br>Concrete blocks consisting of a mixture of volcano ashes and cement were used.<br> |

**Table 1.**
*Fogo Natural Park venue architectural properties [46].*

the structure a colour-like surrounding landscape. The choice of materials also reflects a deep commitment to sustainability. Local materials reduce the carbon footprint associated with transportation and blend harmoniously with the surrounding environment. The building also supports the green building concept by using solar panels on the roofs for sustainable energy. Incorporating solar panels on these roofs further exemplifies the building's commitment to sustainability. This maximises energy efficiency and reduces reliance on non-renewable resources. Rainwater collection systems are another integral component of the building. This reduces the demand for municipal water supplies and promotes self-sufficiency.

By integrating with the topography, maintaining vegetation through green roofs, optimising energy use with solar panels, efficiently managing water resources and choosing local materials, this project exemplifies an essential approach to adapting to natural surroundings and conditions.

### 4.2 National tourist route Trollstigen by Reiulf Ramstad Arkitekter, Norway, 2009

Trollstigen is a scenic route with hairpin bends connecting to a waterfall on the mountain. It is part of the Reinheimen National Park [48]. It is part of a nature-based tourism project that consists of 18 driving routes in Norway [49]. The route was opened in the 1930s and has cultural heritage value, bringing together history, nature, art and architecture [50]. One of the risks of the route is that it is open to natural hazards and only in summer.

The area's natural values are its dramatic landscapes with Nordic fjords; however, this is heavily exploited, and the government and private sectors collaborate to enhance tourism, which results in the built environment expanding into natural areas. On the other hand, considering natural routes in a comprehensive approach and developing a national identity are crucial in planning.

Since the project is tourism oriented and has a dense architectural program with a car park, café and restaurant, sale and exhibition units, WC and technical spaces, it has a significant footprint on the ground in a natural area. It is at risk of expanding further as visitor capacity demands rise and more space is required.

Corten and concrete were selected for the facade to visually adapt to nature, and the sale and exhibition units are half-buried to integrate with the landscape (**Table 2**).

### 4.3 Yellow Submarine Tank by second floor architects, 2015

The project is located at the perimeter of Khao Yai National Park. Since tropical forests and habitats form the area, biodiversity and the chance to encounter wildlife are high [52]. The project area is in a weekend getaway vacation town near the park, but the natural environment continues in the settlements. The visitor rate is high, and the built environment and infrastructure are well-developed, which might result in the region's exploitation and expansion of the built environment.

The project's location at the outer perimeter of the park means general planning policies and settlement decisions were considered. The architectural program and

| **National Tourist Route Trollstigen** | **Visitor centre** | |
|---|---|---|
| Architect: Reiulf Ramstad Architects<br>Total Area: 200000 $m^2$<br>Year: 2010<br>Location: Norway |  |  |
| Architectural Program:<br>It includes the design of a route within the park, including a visitor centre, restaurant, exhibition space, walking paths and viewpoints.<br> | 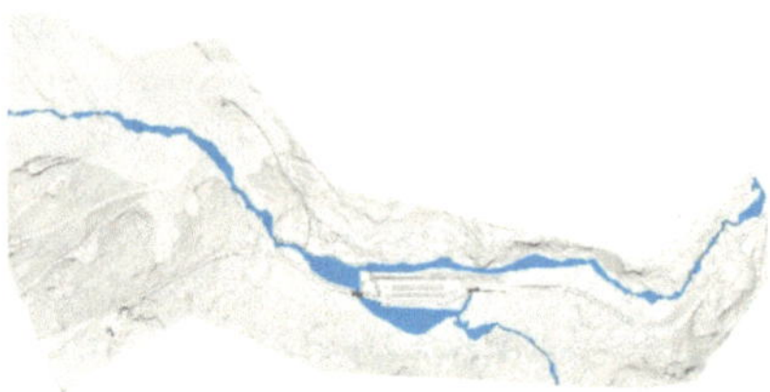 | Design and Spatial Set-Up:<br>The program was designed by combining separate structures. There is a car park on the main road. From there, you can reach the restaurant overlooking the water feature and the rows of exhibition and sales units buried in the ground. |
| 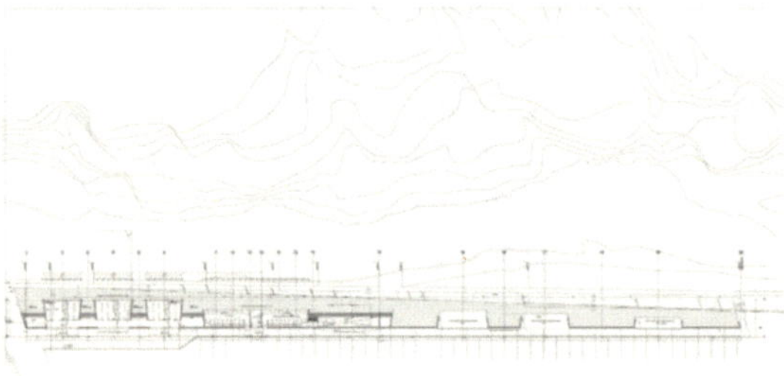 | | 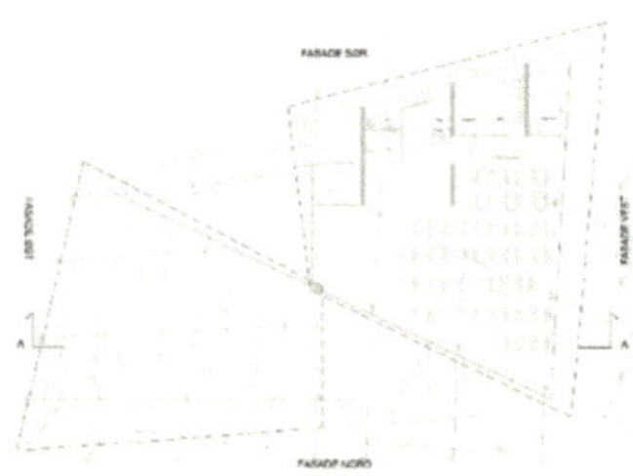 |
| Building and Construction System:<br>While the reinforced concrete shell of the restaurant structure was left without any coating material, the row units in contact with the ground were covered with reinforced concrete carrier corten. | <br> | All bridges, viewing points and landscape elements in the area were designed holistically, considering the harmony of corten and concrete.<br> |

**Table 2.**
*National tourist route Trollstigen architectural properties [51].*

rising visitor demands suggest that the region is at risk of construction. In 2022, a new addition to the café, also selected as the Building of the Year by ArchDaily, was made with a pavilion to rest, play and walk, which proves that.

The project is designed to isolate itself and reflect nature within high walls. Its walls are black and neutral to balance the landscape, and the path to the building and the backyard are covered with gravel. The project aims to sense nature with sounds, shades and touch. The brickwork with gaps creates shades and allows sunlight by play, and the construction site is used mainly as a backyard, with only one of the parts being physically conditioned and consisting of a fully closed space (**Table 3**).

**Yellow Submarine Coffee Tank** — **CAFÉ**

Architect: Second floor Architects
Year: 2015
Total area: 300 m2
Location: Pak Chong, Thailand

Architectural Program:
There is a café and parking lot.
It aims to read the area and its architecture by providing an entrance to the building with a slight slope from the parking lot.

Interaction with Nature:
By isolating nature from the walls surrounding the building, the project aims to notice overlooked elements, from simple human interactions to nature's beauties.

Design and Spatial Set-Up:
Located in a forest area surrounded by a particular tree species, it has an open space surrounded by 300 m2 walls of 3 m on a 1600 m2 land. This area is intended to focus on the inside. The building occupies only one-third of this open space, and visitors are exposed to outdoor conditions in the remaining parts.

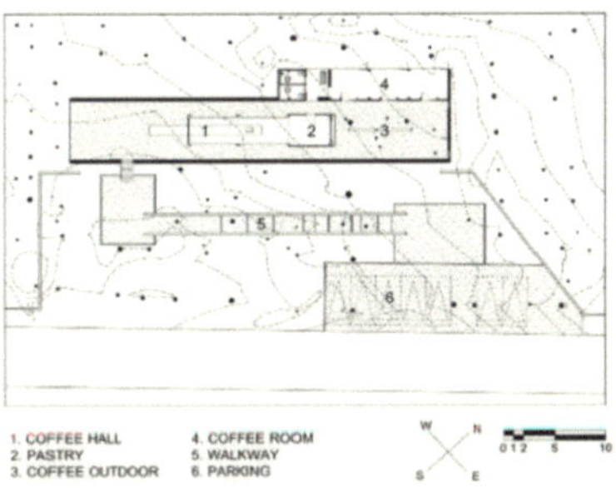


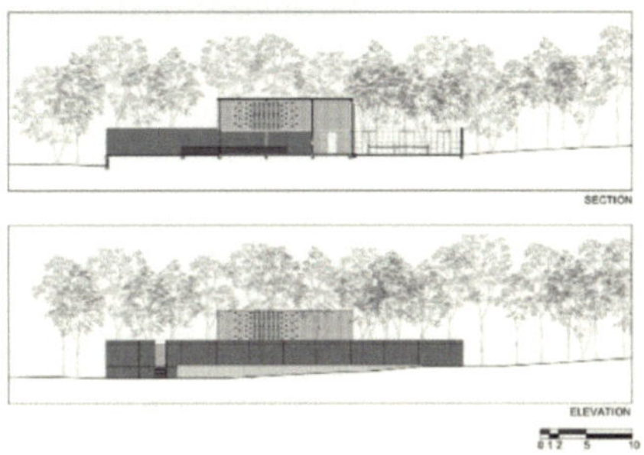


Building and Construction System:
The building has a simple steel construction.

Facade and Shell Design:
The exterior facade is composed of materials in different shades of black, which affect the visibility of some elements through light absorption and reflection. The effects resulting from the physical interactions of different materials have been revealed.

Materials:
Bricks that absorb moisture, footsteps in gravelly outdoor spaces, thick walls that block noise, the reflection of light by exposed concrete walls covered with Chinese ink, the reflection of shiny tiles... In this way, the visibility of black is ensured.

**Table 3.**
*Yellow Submarine Coffee Tank architectural properties [53].*

## 5. Discussion and conclusions

The proposed framework, which includes identifying degradation risks, developing long-term planning policies and establishing architectural design criteria, offers a holistic approach to national park management. This framework is further supported by case studies of award-winning architectural projects that successfully integrate with their natural surroundings. These examples underscore the potential for architecture to harmonise with nature, minimising environmental impacts while enhancing the visitor experience.

The Fogo Natural Park Venue project aimed to harmonise sustainable architecture with the natural volcanic landscape while addressing local community needs and conservation challenges. The building's design successfully integrated with the environment by following the natural topography, using green roofs for vegetation, volcanic ash in construction materials, solar panels for energy efficiency and rainwater collection systems to reduce resource reliance. However, the project lacked sufficient planning for volcanic risks, leading to its destruction by volcanic activity, highlighting the importance of incorporating hazard assessments in future developments. Despite this, the project effectively promoted local involvement and sustainable practices, offering valuable lessons in balancing environmental preservation with human needs.

The Trollstigen Scenic Route project combines nature-based tourism, cultural heritage and architecture within Norway's Reinheimen National Park. While its design integrates with the landscape using corten steel and concrete, and the exhibition units are half-buried to reduce visual impact, the project poses challenges. The dense architectural program, including parking, a café and other facilities, creates a significant footprint in the natural area. As visitor numbers grow, further expansion and environmental degradation are risks. Seasonal access limits environmental strain, but careful planning and sustainable strategies are crucial to balancing tourism with conservation.

The project near Khao Yai National Park blends well with its natural surroundings using high walls, neutral tones and sensory design elements like gravel paths and brickwork with gaps to create shade and natural light. However, its proximity to a popular vacation area and growing visitor demand pose risks of further development, which could threaten the park's biodiversity. While the design respects nature, future expansions—like the 2022 pavilion addition—raise concerns about overdevelopment. Managing growth carefully will protect the natural environment and prevent tourism-driven exploitation.

In conclusion, the study emphasises balancing human activities with preserving natural areas, particularly in national parks. The rising popularity of nature-based tourism presents significant challenges to the sustainability of these protected areas, necessitating a comprehensive approach to their design and management. By analysing the UN's flagship restoration projects and the national park management plans of four diverse countries, this research provides valuable insights into the principles and strategies that can help mitigate the risks of environmental degradation.

The study advocates for a proactive and integrated approach to national park design, where human use and natural preservation are carefully balanced. By prioritising conservation, engaging local communities and adhering to sustainable development practices, the ecological integrity of these vital natural spaces can be protected for future generations. Secondly, buildings cannot be evaluated solely based on their architectural context. Considering degradation and protection probabilities and national parks' management policies are crucial.

## Conflict of interest

The authors declare no conflict of interest.

## Author details

Zeynep Yazicioğlu Halu* and Ayşenur Begüm Helvaci
Istanbul University, Istanbul, Turkey

*Address all correspondence to: zeynep.yaziciogluhalu@istanbul.edu.tr

## References


[1] Kuenzi C, McNeely J. Nature-based tourism. International Risk Governance Council Bookseries [Internet]. 2020;**1**:155-178. DOI: 10.1007%2F978-1-4020-6799-0_8

[2] Butler R. Sustainable tourism in sensitive environments: A wolf in sheep's clothing? Sustainability. 2018;**10**(6):1789

[3] Honey M, Gilpin R, States U. Tourism in the Developing World: Promoting Peace and Reducing Poverty. Washington, D.C.: United States Institute of Peace; 2009

[4] IUCN. Guidelines for Applying Protected Area Management Categories. Switzerland: Gland; 2013

[5] Combining ecological and recreational aspects in national park management: A choice experiment application. Ecological Economics [Internet]. 2011;**70**(6):1231-1239. Available from: https://www.sciencedirect.com/science/article/pii/S092180 0911000607?casa_token=4G08NvgMnFcAAAAA: TDmnvLjU-1CerQEL1PwCJz7-YE8 YVVkl1KVm5_as1nxCoFPvvC7n7Fyo Y2n9obOTWnR9a23T2zY

[6] Moore A. National Parks Planning: A Manual with Annotated Examples. Italy: Food and Agriculture Organization of the United Nations. Available from: https://www.google.com.tr/books/edition/National_Parks_Planning/BSdCVfMxqf4C?hl=en&gbpv=0; 1988, 1976 [Accessed: 9 August 2024]

[7] Agee JK, Johnson DR. Ecosystem Management for Parks and Wilderness. Seattle: University of Washington Press; 1988

[8] Eagles PF, McCool SF. Tourism in National Parks and Protected Areas: Planning and Management. Cabi Publishing; 2002. DOI: 10.1079/9780851995892.0000 [Accessed: 9 August 2024]

[9] Prato T, Fagre D. National Parks and Protected Areas: Approaches for Balancing Social, Economic, and Ecological Values. United Kingdom: Wiley; 2005

[10] Randle EJ, Hoye R. Stakeholder perception of regulating commercial tourism in Victorian National Parks, Australia. Tourism Management. 2016;**54**:138-149. ISSN 0261-5177. Available from: https://www.sciencedirect.com/science/article/pii/S026151771530039X [Accessed: 9 August 2024]

[11] Schwartz Z, Stewart W, Backlund EA. Visitation at capacity-constrained tourism destinations: Exploring revenue management at a national park. Tourism Management. 2012;**33**(3):500-508. Available from: https://www.sciencedirect.com/science/article/pii/S0261517711001105 [Accessed: 9 August 2024]

[12] Stokke KB, Haukeland JV. Balancing tourism development and nature protection across national park borders–a case study of a coastal protected area in Norway. Journal of Environmental Planning and Management. 2018;**61**(12):2151-2165. DOI: 10.1080/09640568.2017.1388772 [Accessed: 9 August 2024]

[13] Pearce J, Dowling R. Monitoring the quality of the visitor experience: An evolutionary journey. Journal of Outdoor Recreation and Tourism. 2019;**25**:87-90. ISSN 2213-0780. Available from: https://www.sciencedirect.com/science/article/

pii/S2213078017300725 [Accessed: 9 August 2024]

[14] Velmurugan S, Thazhathethil BV, George B. A study of visitor impact management practices and satisfaction at Eravikulam National Park, India. International Journal of Geoheritage and Parks. 2021;**9**(4):463-479

[15] Mitchell R, Wooliscroft B, Higham JE. Applying sustainability in national park management: Balancing public and private interests using a sustainable market orientation model. Journal of Sustainable Tourism. 2013;**21**(5):695-715. DOI: 10.1080/09669582.2012.737799 [Accessed: 5 August 2024]

[16] Takyi SA, Seidel AD. Adaptive management in sustainable park planning and management: A case study of the city of Vancouver parks. Journal of Urban Ecology. 2017;**3**(1):juw009

[17] Dressler WH. Co-opting conservation: Migrant resource control and access to national park management in the Philippine uplands. Development and Change. 2006;**37**(2):401-426. DOI: 10.1111/j.0012-155X.2006.00483.x [Accessed: 9 August 2024]

[18] Carter RW, Walsh SJ, Jacobson C, Miller ML. Global change and human impact challenges in managing iconic national parks. In: The George Wright Forum. Vol. 31. Michigan: George Wright Society; 2014. pp. 245-255

[19] Jacobs B, Boronyak L, Mitchell P, Vandenberg M, Batten B. Towards a climate change adaptation strategy for national parks: Adaptive management pathways under dynamic risk. Environmental Science & Policy. 2018;**89**:206-215. Available from: https://www.sciencedirect.com/science/article/pii/S146290111830248X [Accessed: 9 August 2024]

[20] Runyon AN, Carlson AR, Gross J, Lawrence DJ, Schuurman GW. Repeatable approaches to work with scientific uncertainty and advance climate change adaptation in US national parks. In: Parks Stewardship Forum (Vol. 36, No. 1). Berkeley: University of California; 2020

[21] Lindenmayer DB, MacGregor C, Dexter N, Fortescue M, Cochrane P. Booderee National Park management: Connecting science and management. Ecological Management & Restoration. 2013;**14**(1):2-10

[22] Malekpour S, Newig J. Putting adaptive planning into practice: A meta-analysis of current applications. Cities. 2020, 2020;**106**:102866. ISSN 0264-2751. Available from: https://www.sciencedirect.com/science/article/pii/S0264275120312142 [Accessed: 9 August 2024]

[23] Ly TP, Xiao H. The choice of a park management model: A case study of Phong Nha-Ke bang National Park in Vietnam. Tourism Management Perspectives. 2016;**17**:1-15

[24] Van der Merwe H, Bezuidenhout H, Bradshaw PL. Landscape unit concept enabling management of a large conservation area: A case study of Tankwa Karoo National Park, South Africa. South African Journal of Botany. 2015;**99**:44-53. Available from: https://www.sciencedirect.com/science/article/pii/S0254629915002343 [Accessed: 9 August 2024]

[25] Gundersen V, Mehmetoglu M, Vistad OI, Andersen O. Linking visitor motivation with attitude towards management restrictions on use in a national park. Journal of Outdoor Recreation and Tourism. 2015;**9**:77-86. Available from: https://www.sciencedirect.com/science/article/pii/S2213078015000109 [Accessed: 9 August 2024]

[26] Dizdaroğlu D. Developing design criteria for sustainable urban parks. Journal of Contemporary Urban Affairs. 2022;**6**(1):69-81. DOI: 10.25034/ijcua.2022.v6n1-7 [Accessed: 10 August 2024]

[27] Machlis GE, Tichnell DL. The State of the world's Parks: An International Assessment for Resource Management, Policy, and Research. New York: Routledge; 2019

[28] UNEP's Ten Flagship Initiatives Boost Nature and Livelihoods Worldwide [Internet]. Available from: https://www.unep.org/interactive/flagship-initiatives-boosting-nature-livelihoods/#1

[29] About the UN Decade [Internet]. UN Decade on Restoration. Available from: https://www.decadeonrestoration.org/about-un-decade

[30] United Nations. Forests, desertification and biodiversity—United Nations sustainable development [Internet]. In: United Nations Sustainable Development. 2023. Available from: https://www.un.org/sustainabledevelopment/biodiversity/

[31] FAO. Principles for Ecosystem Restoration to Guide the United Nations Decade 2021-2030.2021. Available from: https://openknowledge.fao.org/server/api/core/bitstreams/b234f058-9f77-4481-b870-a7fa2e7ad5f8/content

[32] BRE Global. BREEM Education 2008. 2010

[33] USBGC. The Sustainable Sites Initiative Guidelines and Performance Benchmarks 2009. ASLA, Lady Bird Johnson Wildflower Center at the University of Texas; United States Botanic Garden. ABD: USBGC; 2009

[34] Sustainable Sites Initiative. SITES v2 Rating System for Sustainable Land Design and Development. Austin, TX, USA: Sustainable Sites Initiative; 2014. pp. 1-151

[35] National Park Service. Management Policies 2006. Washington, DC; 2006

[36] South African National Parks. A Framework for Developing and Implementing Management Plans for South African National Parks. 2008

[37] South African National Parks. Kruger National Park Management Plan 2018-2028. 2018

[38] New Zealand Conservation Authority. General Policy for National Parks. Wellington; 2005

[39] Department of Conservation. Tongariro National Park Management Plan. Tongariro; 2006

[40] Ministry of Environment of the Republic of Albania. Management of the Prespa National Park in Albania 2014-2024. Korca; 2013

[41] Texier-Teixeira P, Chouraqui F, Perrillat-Collomb A, Lavigne F, Cadag JR, Grancher D. Reducing volcanic risk on Fogo volcano, Cape Verde, through a participatory approach: Which outcome? Natural Hazards and Earth System Sciences. 2014;**14**(9):2347-2358

[42] BirdLife Data Zone [Internet]. Birdlife.org. 2022 [cited 2024 Aug 11]. Available from: https://datazone.birdlife.org/site/factsheet/6144

[43] Fogo [Internet]. Unesco.org. 2020 [cited 2024 Aug 11]. Available from: https://www.unesco.org/en/mab/fogo

[44] Heart of Fire [Internet]. UDNP Nature Exposure. 2024 [cited 2024 Aug 11]. Available from: https://undp-nature.exposure.co/heart-of-fire

[45] Adrian Kasperski on Coexisting with Nature: The Island of the Future | [Internet]. Archidatum.com. 2014 [cited 2024 Aug 11]. Available from: https://www.archidatum.com/projects/adrian-kasperski-on-coexisting-with-nature-the-island-of-the-future/

[46] Fogo Natural Park Venue / OTO [Internet]. ArchDaily. 2019 [cited 2024 Jul 9]. Available from: https://www.archdaily.com/517828/fogo-national-park-venue-oto-arquitectos?ad_source=myad_bookmarks&ad_medium=bookmark-open

[47] Award-winning building destroyed by volcano [Internet]. Dezeen. 2015. Available from: https://www.dezeen.com/2015/09/28/award-winning-building-destroyed-by-volcano-oto-arquitectos-parque-natural-do-fogo/

[48] Reinheimen National Park Norway's national parks [Internet]. Nasjonalparker. 2024 [cited 2024 Aug 11]. Available from: https://www.norgesnasjonalparker.no/en/nationalparks/reinheimen/

[49] Breiby MA. Exploring aesthetic dimensions in nature-based tourist experiences. Tourism Analysis. 2015;**20**(4):369-380

[50] Halpern N. Accessibility and seasonality of tourism in the Geiranger/Trollstigen area. Working Report M 0705. Molde University College; 2007

[51] National Tourist Route Trollstigen / Reiulf Ramstad Arkitekter + Oslo Norway [Internet]. ArchDaily. 2009 [cited 2024 Jul 9]. Available from: https://www.archdaily.com/32441/national-tourist-route-trollstigen-rra?ad_source=myad_bookmarks&ad_medium=bookmark-open

[52] Discover—Khao Yai National Park [Internet]. Khao Yai National Park. 2021 [cited 2024 Aug 11]. Available from: https://khaoyainationalpark.com/en/discover

[53] Yellow Submarine Coffee Tank / Secondfloor Architects [Internet]. ArchDaily. 2016 [cited 2024 Aug 11]. Available from: https://www.archdaily.com/794208/yellow-submarine-coffee-tank-secondfloor-architects?ad_source=myad_bookmarks&ad_medium=bookmark-open

Chapter 3

# Landscape Dynamics and Biodiversity Conservation in the Ethiopian Highland: Assessing Impacts and Strategies for Sustainability

*Annissa Muhammed and Eyasu Elias*

## Abstract

This study investigates land use and land cover changes (LULCC) in the Bale Mountains National Park (BMNP) from 1985 to 2017, highlighting significant reductions in forestland, ericaceous land, and grassland, with a marked increase in agricultural areas. The landscape metrics revealed increased habitat fragmentation and edge effects, negatively impacting biodiversity and ecosystem functions. Five distinct plant community types were identified, demonstrating varying levels of species richness and diversity. Soil analyses indicated degradation in edge habitats compared to interior areas. The study underscores the need for robust conservation strategies, including enhanced forest protection, land restoration, sustainable agriculture, and long-term habitat monitoring. Recommendations emphasize community involvement, soil conservation, and policy support to mitigate LULCC impacts and promote sustainable management of the BMNP, ensuring the preservation of its unique biodiversity and ecological integrity. These measures aim to balance conservation with the socio-economic needs of local communities.

**Keywords:** biodiversity, ecological integrity, habitat fragmentation, landscape metrics, sustainable management

## 1. Introduction

Ethiopia is renowned for its diverse ecological conditions, encompassing a wide range of climates and habitats, from lowland deserts to highland forests. This ecological diversity supports a rich array of flora and fauna, many of which are endemic to the region. The Bale Mountains National Park (BMNP) is a critical ecological zone, hosting a variety of ecosystems including tropical montane forests, ericaceous shrublands, afro-alpine meadows, and grasslands [1, 2]. These ecosystems are home to numerous endemic species, making the BMNP a biodiversity hotspot of global significance.

Despite its ecological importance, the BMNP faces significant threats from anthropogenic activities. Over the past few decades, the park has experienced extensive land use and land cover changes (LULCC) driven by agricultural expansion, settlement growth, and resource extraction [3, 4]. These activities have resulted in habitat fragmentation, loss of biodiversity, and alterations in ecosystem functions [5]. Similar to the findings in the Canary Archipelago and Madeira Island, where land use changes have led to substantial shifts in land cover [6, 7], BMNP has seen significant conversion of forestland into farmland and other land cover types. This conversion has led to the degradation of critical habitats, threatening the ecological integrity of the park [8] and reflecting broader trends observed in other insular urban territories undergoing similar land use dynamics [6]. The extensive deforestation and habitat degradation are primarily driven by human activities, similar to other ecological zones worldwide, emphasizing the urgent need for effective land management and conservation strategies.

Understanding the precise extent and impact of LULCC in BMNP is crucial for developing effective conservation strategies. However, there is a significant research gap in this area. Previous studies have primarily provided general descriptions of land cover changes without detailed quantifications and analyses of the transitions between different land cover types. Additionally, comprehensive studies examining the relationships between landscape pattern metrics and their effects on vegetation composition and structure are lacking. The interactions between soil properties and habitat gradients in the context of LULCC also remain underexplored. Addressing these gaps is essential for formulating informed and effective conservation policies.

This study is of paramount importance for several reasons. First, it offers a detailed and quantitative analysis of land cover changes in BMNP over three decades, providing valuable insights into the trends and trajectories of these changes. Understanding these patterns is essential for identifying the drivers of habitat loss and prioritizing conservation efforts. Second, the study analyzes the correlations among various landscape pattern metrics, aiding in understanding the spatial configuration and fragmentation of habitats. Such information is critical for designing landscapes that maintain ecological connectivity and resilience.

Moreover, the study investigates the effects of landscape pattern changes on vegetation composition and structure, offering insights into how different land cover types influence biodiversity. This is particularly important for BMNP, where maintaining species diversity is crucial for ecosystem stability and functioning. Additionally, the analysis of soil physical and chemical properties across habitat gradients provides a deeper understanding of how LULCC affects soil health and productivity. This knowledge is essential for promoting sustainable land use practices that preserve soil quality. The overall objective of this study is to comprehensively assess the impact of LULCC on the ecological conditions of the BMNP over the past three decades, with a focus on quantifying land cover transformations, understanding habitat fragmentation and ecological connectivity, evaluating changes in vegetation composition and structure, and investigating the associated effects on soil physical and chemical properties.

## 2. Methodology

### 2.1 Study area description

Bale Mountain National Park (BMNP) is located in the Oromia Region of southeastern Ethiopia, approximately 400 kilometers southeast of Addis Ababa. The park, which spans an area of about 2200 square kilometers, lies between latitudes 6°29′–7°10′N and longitudes 39°28′–39°57′E (**Figure 1**) [9]. The park encompasses diverse vegetation types, ranging from grasslands to high-altitude Afro-alpine ecosystems. Key vegetation zones include the Harenna Forest, a montane forest in the southern part of the park that is one of Ethiopia's largest intact forests, rich in biodiversity. The park also features Afro-montane forests dominated by *Hagenia abyssinica*, *Juniperus procera*, *Hypericum revolutum*, and *Afro-alpine moorlands* characterized by giant lobelias (Lobelia rhynchopetalum) and Erica species [10, 11]. The park experiences varied agro-climatic conditions due to its significant altitudinal range, which spans from around 1500 meters to 4377 meters at the summit of Tullu Demtu, Ethiopia's second-highest peak. This variation creates diverse habitats and microclimates. The climate is generally characterized by a wet season from March to October and a dry season from November to February, with average annual rainfall ranging from 1000 to 1400 mm [12]. BMNP's soils are highly variable, influenced by the underlying geology and

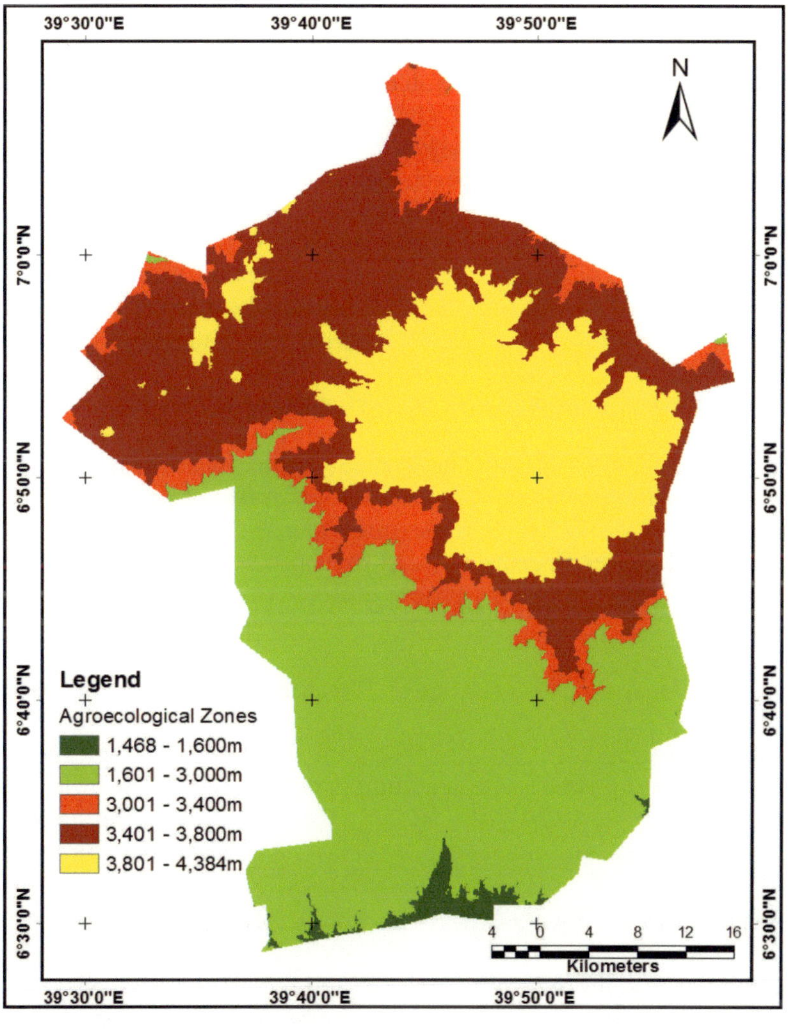


**Figure 1.**
*Map of the BMNP showing agroecological zones.*

topography. Key soil types include Andosols in the higher altitudes, known for their fertility and good drainage; Nitisols in lower elevations, which are deep and fertile; and Vertisols in valley bottoms, which are clay-rich and prone to waterlogging during the rainy season but fertile when managed properly [13, 14].

### 2.2 Land cover change detection

To analyze the land cover changes in the BMNP, we utilized remote sensing data and geographic information system (GIS) techniques [15]. Land cover maps from 1985 to 2017 were generated using satellite images obtained from Landsat and Sentinel missions [16]. These images were classified into different land cover types, including forestland, ericaceous land, grassland, afro-alpine land, and farmland, using supervised classification techniques [17]. Classification accuracy was validated using ground truth data collected through field surveys [18]. Change detection analysis was performed by comparing the land cover maps over different time periods, quantifying the gains, losses, and conversions between land cover types, which were represented in a transition matrix (**Table 1**) [19].

### 2.3 Correlations among landscape pattern metrics

We employed linear regression analysis to explore the relationships among six landscape pattern metrics: mean patch area (AREA_MN), mean shape index (SHAPE_MN), edge density (ED), core area (COA), mean nearest neighbor distance (ENN_MN), and interspersion and juxtaposition index (IJI) [20]. The correlations of these metrics with patch number (PN) were examined to understand the fragmentation and connectivity of the landscape [21]. Additionally, interrelationships among the metrics were assessed to identify significant patterns and trends [22].

### 2.4 Vegetation classification and species composition

Vegetation classification was conducted using hierarchical cluster analysis at an 81-dissimilarity level, resulting in the identification of five plant community types [23]. Dominant species combinations were used to name each community type. Species richness, Shannon diversity index, β-diversity, and Pielou's evenness index were calculated for each community type to evaluate the floristic composition and structure [24].

### 2.5 Effects of landscape pattern change on floristic composition and structure

We performed regression analysis to determine the effects of landscape pattern changes on species composition and structural parameters [23]. The influence of landscape metrics such as PN, AREA_MN, SHAPE_AM, COA, and ED on species richness, Shannon diversity index, and woody species density and basal area was assessed [21].

### 2.6 Soil physical and chemical properties across habitat gradient

Soil samples were collected from both edge and interior habitats of the BMNP. Physical properties such as sand, silt, and clay contents, bulk density (BD), and the

**(a) 1985–1995**

| | | | | | 1985 | | |
|---|---|---|---|---|---|---|---|
| | **LULC Types** | **Gr** | **Af** | **Fo** | **Er** | **Total** | **Gain** |
| 1995 | Grassland (Gr) | **18,703** | 3416 | 328 | 5039 | 27,485 | 8782 |
| | Afro-alpine (Af) | 5406 | **21,560** | 3 | 2382 | 29,351 | 7791 |
| | Forestland (Fo) | 1887 | 8 | **83,263** | 2067 | 87,225 | 3962 |
| | Erica (Er) | 1530 | 4360 | 3600 | **64,226** | 73,716 | 9490 |
| Total | | 27,525 | 29,345 | 87,195 | 73,712 | 217,777 | |
| Loss | | 8822 | 7784 | 3931 | 9487 | | |

**(b) 1995–2005**

| | | 1995 | | | | | |
|---|---|---|---|---|---|---|---|
| | **LULC Types** | **Gr** | **Af** | **Fo** | **Er** | **Total** | **Gain** |
| 2005 | Grassland (Gr) | **19,060** | 5745 | 77 | 1380 | 26,262 | 7202 |
| | Afro-alpine (Af) | 1481 | **19,317** | 3421 | 5124 | 29,343 | 10,026 |
| | Forestland (Fo) | 360 | 9 | **82,099** | 4707 | 87,174 | 5075 |
| | Erica (Er) | 5905 | 4262 | 1002 | **62,363** | 73,532 | 11,169 |
| | Farmland (Fa) | 680 | 18 | 626 | 144 | 1467 | 1467 |
| Total | | 27,485 | 29,351 | 87,225 | 73,716 | 217,777 | |
| Loss | | 8425 | 10,034 | 5126 | 11,354 | | |

**(c) 2005–2017**

| | | | | | 2005 | | | |
|---|---|---|---|---|---|---|---|---|
| | **LULC Types** | **Gr** | **Af** | **Fo** | **Er** | **Fa** | **Total** | **Gain** |
| 2017 | Grassland (Gr) | **17,157** | 4058 | 63 | 3902 | 40 | 25,170 | 8063 |
| | Afro-alpine (Af) | 6978 | **17,301** | 7 | 4846 | 209 | 29,331 | 12,040 |
| | Forestland (Fo) | 40 | 6 | **81,709** | 5103 | 286 | 87,034 | 5435 |
| | Erica (Er) | 1793 | 7914 | 4258 | **59,297** | 207 | 73,518 | 14,172 |
| | Farmland (Fa) | 295 | 14 | 1188 | 384 | **725** | 2725 | 1880 |
| Total | | 26,262 | 29,343 | 87,174 | 73,532 | 1467 | 217,777 | |
| Loss | | 9106 | 11,993 | 5516 | 14,235 | 743 | | |

| **(d) 1985–2017** | | | | | | | | |
|---|---|---|---|---|---|---|---|---|
| | **LULC Types** | **Gr** | **Af** | **Fo** | **Er** | **Total** | **Gain** |
| 2017 | Grassland (Gr) | **16,322** | 5526 | 269 | 3053 | 25,170 | 8848 |
| | Afro-alpine (Af) | 6255 | **18,702** | 6 | 4378 | 29,331 | 10,639 |
| | Forestland (Fo) | 163 | 15 | **79,381** | 7585 | 87,034 | 7763 |
| | Erica (Er) | 4220 | 5095 | 6297 | **57,906** | 73,518 | 15,612 |
| | Farmland (Fa) | 565 | 7 | 1243 | 789 | 2725 | 2725 |
| Total | | 27,525 | 29,345 | 87,195 | 73,712 | 217,777 | |
| Loss | | 11,203 | 10,643 | 7815 | 15,805 | | |

**Table 1.**
*The dynamics of land cover changes from 1985 to 2017 (area in ha).*

Si/Cl ratio were measured [25]. Chemical properties including total nitrogen (TN), available phosphorus (AP), carbon to nitrogen (C/N) ratio, soil organic carbon (SOC), cation exchange capacity (CEC), and pH were analyzed [26]. The differences between the edge and interior habitats were statistically tested to determine the impact of habitat gradient on soil properties [27].

## 3. Result

### 3.1 Land cover change detection

**Table 1** and **Figure 2** illustrate the dynamics of land cover changes in the BMNP over time. The major diagonals represent the areas of each land cover type that remained unchanged, while the off-diagonals indicate the gains, losses, and trajectories of conversions between land cover types. For instance, **Table 1(d)** shows that out of 87,197 hectares of forestland in 1985, 79,381 hectares remained stable, while 6293 hectares were converted to ericaceous land, 1243 hectares to farmland, 269 hectares to grassland, and 6 hectares to afro-alpine land. This results in total forestland losses of 7815 hectares, with an annual conversion rate of 1.65 hectares. Conversely, some of the 87,034 hectares of forestland in 2017 were gained from other land cover types: 7585 hectares from ericaceous land, 163 hectares from grassland, and 15 hectares from afro-alpine land, resulting in total gains of 7763 hectares.

### 3.2 The correlations among the landscape pattern metrics

Based on the linear regression analysis, six landscape metrics—AREA_MN, SHAPE_MN, ED, COA, ENN_MN, and IJI—showed a significant relationship with PN ($p < 0.05$) (**Figure 3**). Additionally, COA and IJI were significantly related to AREA_MN, and ENN_MN was significantly related to ED ($p < 0.05$). Metrics positively correlated with PN included SHAPE_MN ($r = 0.58$, $p < 0.05$), ED ($r = 0.87$, $p < 0.001$), and IJI ($r = 0.58$, $p < 0.05$). Conversely, metrics negatively correlated

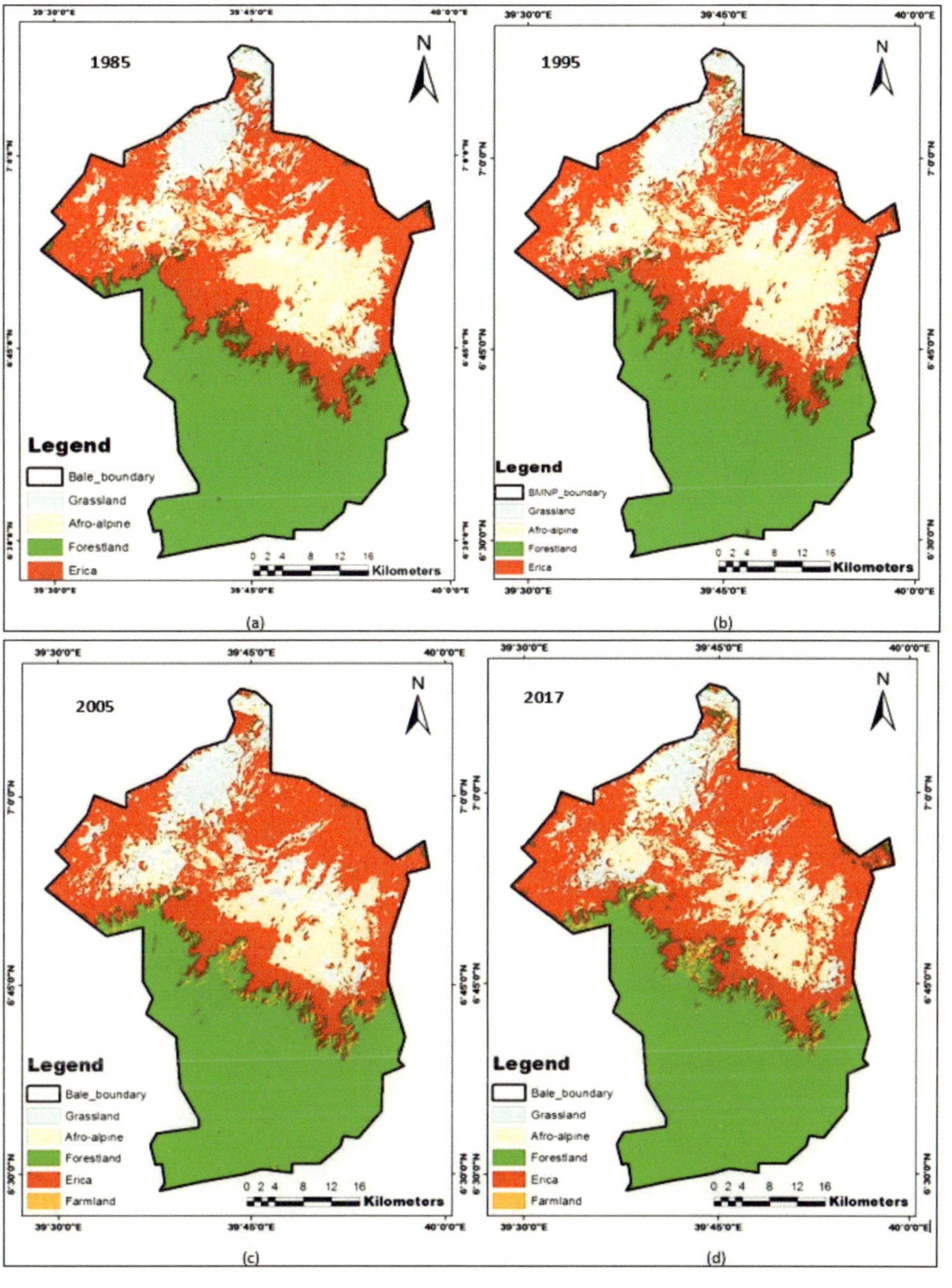


**Figure 2.**
*LULCC map of BMNP for the year 1985 (a), 1995 (b), 2005 (c), and 2017 (d). (Source: Muhammed and Elias [9]).*

with PN included AREA_MN ($r = -0.71$, $p < 0.01$), COA ($r = -0.64$, $p < 0.05$), and ENN_MN ($r = -0.77$, $p < 0.01$).

Furthermore, AREA_MN showed a positive correlation with COA ($r = 0.58$, $p < 0.05$) and negative correlations with ENN_MN ($r = -0.91$, $p < 0.001$) and IJI ($r = -0.58$, $p < 0.05$). COA demonstrated negative correlations with ENN_MN

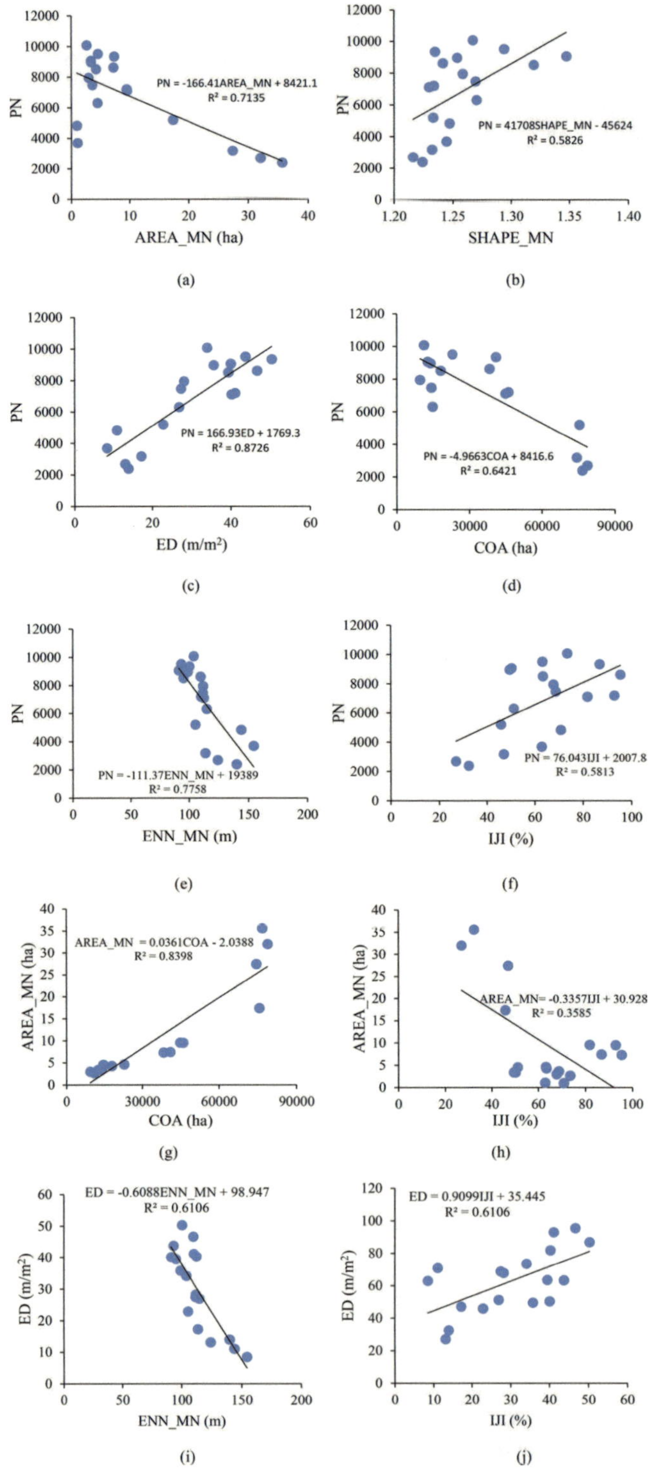


**Figure 3.**
*Correlation among PN and AREA_MN (a), PN and SHAPE_MN (b), PN and ED (c), PN and COA (d), PN and ENN_MN (e) , PN and IJI (f), AREA_MN and COA (g), AREA_MN and IJI (h), ED and ENN_MN (i), ED and IJI (j).*

($r = -0.81$, $p < 0.01$) and IJI ($r = -0.61$, $p < 0.05$). ED was positively associated with IJI ($r = 0.57$, $p < 0.05$) and negatively associated with ENN_MN ($r = -0.61$, $p < 0.05$) (**Figure 3**). These relationships indicate high predictability of these variables based on the others.

The landscape of the BMNP was characterized as moderately fragmented, indicated by increased values of PN, SHAPE_MN, ED, ENN_MN, and IJI, and decreased values of AREA_MN and COA. This fragmentation is attributed to human encroachment and resultant LULCC.

### 3.3 Vegetation classification

The vegetation classification at 81-dissimilarity level from hierarchical cluster analysis resulted in five plant community types. The name of each community type was given after two or more dominant species combinations that have higher synoptic values. The five community types identified were *Anthemis tigreensis – Alchemilla pedata* represented as community type one (C1), *Alchemilla haumanii – Helichrysum gofense* as community type two (C2), *Podocarpus falcatus – Croton macrostachyus* as community type three (C3), *Croton macrostachyus – Syzygium guineense – Olea capensis* as community type four (C4), and *Solanum marginatum – Euphorbia dumalis – Rubus steudneri* as community type five (C5). Three classification levels were distinguished in the cluster dendrogram. In the first hierarchy, C5 was separated from C1, C2, C3, and C4; in the second hierarchy C1 and C2 were separated from C3 and C4; in the third hierarchy C1 was separated from C2; and C3 was separated from C4.

### 3.4 Species composition in the plant communities

The result of this study revealed that the highest species richness, Shannon, and ß-diversity were obtained in community type 1 that extends from 3000 m to 3400 m asl and constituted woodland and grassland vegetation. Conversely, the least species richness, Shannon and ß-diversity were scored by community type 5 which is laid from 3000 m to 3200 m asl and mainly comprised of vegetation in the farmland and settlements. In addition, community type 5 showed the highest Pielou's evenness index, whereas community type 4 depicted the least (**Table 2**).

### 3.5 Effects of landscape pattern change on floristic composition and structure

The computed regression analysis among the landscape indices and the species composition and structural parameters revealed that PN, AREA_MN, SHAPE_AM,

| Communities | Altitude (m) | S | H′ | J | ßw |
|---|---|---|---|---|---|
| C1 | 3000–3400 | 90 | 3.77 | 0.84 | 5.35 |
| C2 | 3800–4200 | 43 | 2.93 | 0.78 | 2.91 |
| C3 | 1800–3000 | 47 | 3.00 | 0.78 | 3.12 |
| C4 | 1600–1800 | 52 | 3.00 | 0.76 | 4.30 |
| C5 | 3000–3200 | 23 | 2.78 | 0.89 | 1.34 |

**Table 2.**
*Species richness, diversity, and evenness of the woody communities in BMNP.*

| | | | | | | | | | | | |
|---|---|---|---|---|---|---|---|---|---|---|---|
| **S** | **1** | | | | | | | | | | |
| H′ | 0.95$^{***}$ | 1 | | | | | | | | | |
| BA | 0.96$^{***}$ | 0.94$^{***}$ | 1 | | | | | | | | |
| Density | 0.99$^{***}$ | 0.90$^{***}$ | 0.93$^{***}$ | 1 | | | | | | | |
| PN | −0.90$^{***}$ | −0.96$^{***}$ | −0.96$^{***}$ | −0.84$^{**}$ | 1 | | | | | | |
| AREA_MN | 0.95$^{***}$ | 0.87$^{***}$ | 0.82$^{**}$ | 0.72$^{**}$ | −0.71$^{**}$ | 1 | | | | | |
| SHAPE_AM | −0.42 | −0.64$^{*}$ | −0.53$^{*}$ | −0.28 | 0.48 | −0.18 | 1 | | | | |
| COA | −0.56$^{*}$ | −0.37 | 0.34 | 0.66$^{*}$ | −0.64$^{*}$ | 0.58$^{*}$ | −0.79$^{**}$ | 1 | | | |
| ED | 0.65$^{*}$ | 0.78$^{**}$ | −0.54$^{*}$ | −0.77$^{**}$ | 0.87$^{***}$ | −0.91$^{***}$ | 0.45 | −0.16 | 1 | | |
| ENN_MN | −0.37 | −0.56$^{*}$ | −0.49 | −0.28 | −0.77$^{**}$ | −0.26 | 0.73$^{**}$ | −0.81$^{**}$ | −0.61$^{*}$ | 1 | |
| IJI | −0.42 | −0.28 | −0.32 | −0.37 | 0.58$^{*}$ | −0.59$^{*}$ | 0.37 | −0.61$^{*}$ | 0.57$^{*}$ | 0.17 | 1 |
| | S | H′ | BA | Density | PN | AREA_MN | SHAPE_AM | COA | ED | ENN_MN | IJI |

*Note: S = species richness, H′ = Shannon diversity index, BA = basal area, PN = patch number, AREA_MN = mean patch size, SHAPE_AM = area weighted mean shape index, COA = core area, ED = edge density, ENN_MN = mean Euclidean nearest neighbor distance, IJI = interspersion and juxtaposition index.*

**Table 3.**
*Pearson's correlation among fragmentation indices and floristic composition and structural parameters.*

COA, and ED significantly affected the species composition and structure of the study area. Accordingly, PN was strong and negatively affected the overall species richness (r = −0.90, $p < 0.001$) and Shannon diversity index (r = −0.96, $p < 0.001$) of the study area. Conversely, AREA_MN was strong and positively correlated with the overall species richness (r = 0.95, $p < 0.001$) and Shannon diversity index (r = 0.87, $p = 0.01$) of the study area. Moreover, the effect of ED on species richness (r = 0.65, $p = 0.05$) and Shannon diversity index (r = 0.78, $p = 0.01$) was significant and positive. Whereas, COA and SHAPE_AM were significant and negatively affected species richness (r = −0.56, $p = 0.05$) and Shannon diversity index (r = −0.64, $p = 0.05$), respectively. Moreover, Pielou's evenness index was insignificant and positively correlated with AREA_MN and COA but it was insignificant and negatively correlated with the rest of the landscape structural indices. The effect of landscape structural change on β-diversity (species turnover) was just the opposite of COA. Besides, PN and ED were significant and negatively affected woody species density (r = −0.84, $p < 0.01$ and r = −0.77, $p < 0.01$, respectively) and basal area (r = −0.96, $p < 0.001$ and r = −0.54, $p < 0.05$, respectively). Whereas, the effect of AREA_MN on woody species density (r = 0.72, $p < 0.01$) and basal area (r = 0.82, $p < 0.01$) was significant and positive. Conversely, COA was significant and positively affected woody species density (r = 0.66, $p < 0.05$). In addition, SHAPE_AM was significant and negatively influenced basal area (r = −0.53, $p < 0.05$). However, the woody species density and basal area were insignificant and negatively correlated with ENN_MN and IJI (**Table 3**).

### 3.6 Soil physical and chemical properties across the habitat gradient

The edge and interior habitat significantly ($p < 0.05$) affected some of the soil's physical and chemical properties (**Table 4**). The mean values of sand (55.43 ± 1.04),

| Soil properties | Habitat types | | Mean | *F* | Sig |
|---|---|---|---|---|---|
| | Edge habitat | Interior habitat | | | |
| Sand (%) | 55.43 ± 1.04 | 54.52 ± 0.99 | 54.97 ± 0.71 | 1.45 | ns |
| Silt (%) | 25.39 ± 1.34 | 24.79 ± 1.20 | 25.09 ± 0.89 | 1.04 | ns |
| Clay (%) | 19.18 ± 1.97 | 20.69 ± 1.57 | 19.94 ± 1.25 | 0.76 | ns |
| Si/Cl | 2.31 ± 0.57 | 1.44 ± 0.15 | 1.88 ± 0.30 | 0.082 | ns |
| BD ($g/cm^3$) | 0.75 ± 0.08$^a$ | 0.64 ± 0.06$^b$ | 0.69 ± 0.05 | 8.09 | * |
| pH ($H_20$) | 5.58 ± 0.11 | 5.62 ± 0.11 | 5.60 ± 0.08 | 1.04 | ns |
| SOC (%) | 6.40 ± 0.38 | 6.63 ± 0.39 | 6.51 ± 0.27 | 0.57 | ns |
| TN (%) | 0.56 ± 0.08$^b$ | 0.73 ± 0.08$^a$ | 0.64 ± 0.06 | 15.90 | * |
| C/N ratio | 14.65 ± 1.59$^a$ | 10.67 ± 0.74$^b$ | 12.65 ± 0.91 | 6.35 | * |
| AP (mg/kg) | 9.82 ± 1.05$^b$ | 12.81 ± 1.59$^a$ | 11.31 ± 0.97 | 14.69 | ** |
| CEC (cmol/kg) | 20.19 ± 1.49 | 23.67 ± 1.62 | 21.93 ± 1.12 | 0.76 | ns |

*Note: Means in each column with the same superscript letter are not significantly different at P ≤ 0.05; BD = bulk density, SOC = organic carbon, TN = total nitrogen, C/N = carbon to nitrogen ratio, AP = available phosphorus, CEC = cation exchange capacity, ns = nonsignificant; **p < 0.01 and *p < 0.05.*

**Table 4.**
*The mean (± SE) values of selected soil physical properties as affected by habitat types. BD: Bulk density.*

silt (25.39 ± 1.34) and clay contents (19.18 ± 1.97), as well as the Si/Cl ratio (2.31 ± 0.57) in the soils of edge habitat were not significantly different compared to the mean sand (54.52 ± 0.99), silt (24.79 ± 1.20) and clay (20.69 ± 1.57) contents, as well as Si/Cl ratio (1.44 ± 0.15) in the soils of the interior habitat. However, the mean BD of the soil in the edge habitat (0.75 ± 0.08) was significantly higher compared to the soils in the interior habitat (0.64 ± 0.06). Conversely, the contents of TN (0.56 ± 0.08), AP (9.82 ± 1.05), and C/N ratio (14.65 ± 1.59) in the edge habitat were significantly different compared to the contents of TN (0.73 ± 0.08), AP (12.81 ± 1.59), and C/N ratio (10.67 ± 0.74) in the soils of interior habitat. However, the mean pH, SOC, and CEC of the soil in the edge and interior habitat did not show significant difference. Whereas, the mean pH (5.62 ± 0.11), CEC (23.67 ± 1.62), and SOC (6.63 ± 0.39) were markedly higher in the interior habitat than the mean pH (5.58 ± 0.11), CEC (20.19 ± 1.49), and SOC (6.40 ± 0.38) in the edge habitat.

## 4. Discussion

The results of this study reveal significant LULCC in the BMNP over the past three decades, reflecting broader trends observed in similar ecological zones. This pattern of LULCC is consistent with findings from other regions, such as the Canary Archipelago and Madeira Island, where similar retrospective analyses using CORINE data have shown substantial shifts in land use [6, 7]. In BMNP, the conversion of forestland to agricultural fields and other land uses is particularly concerning, as it threatens the ecological integrity of this biodiversity hotspot. Between 1985 and 2017, there was a substantial decline in forestland, ericaceous land, and grassland, with a concurrent increase in farmland. This pattern of deforestation and habitat degradation is driven by expanding agricultural activities and settlement growth, mirroring trends observed in other insular urban territories where land use changes have been rigorously documented [6].

The most dominant land cover type in 1985, forestland, experienced significant losses, primarily converting to ericaceous land, farmland, and grassland. This aligns with studies that highlight widespread deforestation and habitat degradation due to human activities [7]. The shift in land use patterns in BMNP is indicative of broader socio-economic pressures that drive land cover changes in ecological zones across the globe. The findings emphasize the urgent need for effective land management strategies to mitigate the impacts of LULCC and preserve the ecological integrity of BMNP, much like the measures recommended for other regions experiencing similar land use dynamics [6].

The regression analysis indicated significant relationships between various landscape metrics and patch number (PN), demonstrating the impact of LULCC on habitat fragmentation. Metrics such as AREA_MN, SHAPE_MN, ED, COA, ENN_MN, and IJI showed significant correlations with PN, reflecting changes in habitat structure and connectivity. The positive correlation of SHAPE_MN, ED, and IJI with PN suggests increased landscape complexity and edge effects, while negative correlations with AREA_MN, COA, and ENN_MN indicate reduced core habitat areas and greater isolation of habitat patches [20].

The vegetation classification identified five distinct plant community types, with notable differences in species richness and diversity among them. Community type 1, extending from 3000 to 3400 meters above sea level, exhibited the highest species richness and diversity, likely due to its heterogeneous habitat structure and lower levels of human disturbance [1, 2]. In contrast, community type 5, located at lower elevations and more impacted by agricultural activities, showed the lowest species richness and diversity, highlighting the adverse effects of LULCC on biodiversity [3, 4].

The study also found significant differences in soil properties between edge and interior habitats, emphasizing the impact of LULCC on soil health. Edge habitats exhibited higher bulk density and lower total nitrogen, available phosphorus, and C/N ratios compared to interior habitats, indicating soil compaction and nutrient depletion due to increased human activities [4, 5]. These findings underscore the need for sustainable land management practices to mitigate soil degradation and maintain ecosystem functions.

The observed LULCC and associated ecological impacts in the BMNP underscore the urgent need for effective conservation strategies. Protecting remaining forest areas, restoring degraded lands, and promoting sustainable agricultural practices are crucial to preserving the park's biodiversity and ecological integrity [3, 5]. Integrating community-based conservation approaches can enhance local engagement and ensure the long-term success of conservation efforts [1, 2].

## 5. Conclusion and recommendation

The study reveals significant land cover changes and landscape fragmentation in the BMNP region, driven primarily by human encroachment. The findings indicate considerable conversions between various land cover types from 1985 to 2017, with forestland experiencing both losses and gains. Landscape metrics such as AREA_MN, SHAPE_MN, ED, COA, ENN_MN, and IJI show significant relationships with floristic composition and structure. Community type 1, characterized by woodland and grassland vegetation, exhibited the highest species richness and diversity, while community type 5, mainly comprising farmland and settlements, had the lowest. Soil properties across habitat gradients also differed notably, with edge habitats displaying higher bulk density and lower nutrient contents compared to interior habitats.

To address these challenges, targeted conservation strategies are essential to mitigate human encroachment and preserve stable forestland. Promoting sustainable land use practices, such as agroforestry, can balance agricultural needs with conservation goals. Habitat restoration efforts, particularly in degraded areas, will enhance ecological resilience. Continuous monitoring and adaptive management strategies are crucial for responding to emerging threats. Engaging local communities in conservation efforts and further research on landscape-ecology relationships can inform policy development and promote sustainable land management practices. By implementing these recommendations, the BMNP region can achieve a balanced coexistence between human activities and natural ecosystems, ensuring long-term biodiversity conservation and landscape integrity.

## Author details

Annissa Muhammed[1*] and Eyasu Elias[2]

1 Department of Natural Resource Management, College of Agriculture and Environmental Science, Arsi University, Assela, Ethiopia

2 Centre for Environmental Science, College of Natural and Computational Science, Addis Ababa University, Addis Ababa, Ethiopia

*Address all correspondence to: annissamuhammed@gmail.com

## References

[1] Hillman JC. The Bale Mountains National Park area, Southeast Ethiopia, and its management. Mountain Research and Development. 1988;**8**(2/3):253-258. DOI: 10.2307/3673421

[2] Sillero-Zubiri C. Ecological corridors in the Bale Mountains. Mountain Research and Development. 2008;**28**(3/4):256-261. DOI: 10.1659/mrd.1038

[3] Fetene A et al. Conservation of the Afromontane forest in the Bale Mountains National Park. Environmental Conservation. 2006;**33**(4):340-348. DOI: 10.1017/S0376892906003536

[4] Tesfaye G et al. Socioeconomic factors affecting land use change in Bale Mountains National Park. Environmental Management. 2014;**53**(2):396-409. DOI: 10.1007/s00267-013-0224-6

[5] Mekuria W et al. Restoration of degraded lands in Bale Mountains National Park. Ecological Engineering. 2009;**35**(4):945-952. DOI: 10.1016/j.ecoleng.2008.12.005

[6] Lousada S, Cabezas J, Castanho RA, Gómez JMN. Land-use changes in insular urban territories: A retrospective analysis from 1990 to 2018. The case of Madeira Island—Ribeira Brava. Sustainability. 2022;**14**:16839. DOI: 10.3390/su142416839

[7] Naranjo Gómez JM, Lousada S, Velarde JG, Castanho RA, Loures L. Land-use changes in the canary archipelago using the CORINE data: A retrospective analysis. Land. 2020;**9**:232. DOI: 10.3390/land9070232

[8] Kidane Y et al. Assessing land cover dynamics in Bale Mountains National Park. Journal of Land Use Science. 2012;**7**(1):25-42. DOI: 10.1080/1747423X.2010.520341

[9] Muhammed E, Elias E. Class and landscape level habitat fragmentation analysis in the Bale Mountains National Park. Heliyon. 2021;**7**(3):e06247. DOI: 10.1016/j.heliyon.2021.e06247

[10] Friis I. The forest vegetation of Ethiopia. Symbolae Botanicae Upsalienses. 1986;**26**(2):31-47

[11] Largen MJ, Spawls S. The Amphibians and Reptiles of Ethiopia and Eritrea. Frankfurt am Main, Germany: Edition Chimaira; 2010

[12] Hillman JC. Conservation in the Bale Mountains. Oryx. 1986;**20**(2):89-94. DOI: 10.1017/S0030605300026212

[13] Abebe M, Kifle M. Soil types in Bale Mountains National Park and their agricultural potential. Journal of Soil Science and Environmental Management. 2011;**2**(5):102-110. DOI: 10.5897/JSSEM11.035

[14] Tamrat B. Vegetation of the Bale Mountains, Ethiopia. Mountain Research and Development. 1993;**13**(3):265-274. DOI: 10.2307/3673665

[15] Jensen JR. Introductory Digital Image Processing: A Remote Sensing Perspective. 3rd ed. Upper Saddle River, NJ, USA: Prentice Hall; 2005

[16] Roy DP et al. Landsat-8: Science and product vision for terrestrial global change research. Remote Sensing of Environment. 2014;**145**:154-172. DOI: 10.1016/j.rse.2014.02.001

[17] Lillesand TM, Kiefer RW, Chipman JW. Remote Sensing and Image

Interpretation. 7th ed. Hoboken, NJ, USA: Wiley; 2015

[18] Foody GM. Status of land cover classification accuracy assessment. Remote Sensing of Environment. 2002;**80**(1):185-201. DOI: 10.1016/S0034-4257(01)00295-4

[19] Singh A. Digital change detection techniques using remotely-sensed data. International Journal of Remote Sensing. 1989;**10**(6):989-1003. DOI: 10.1080/01431168908903939

[20] McGarigal K, Marks BJ. FRAGSTATS: Spatial Pattern Analysis Program for Quantifying Landscape Structure. General Technical Report PNW-GTR-351. Portland, OR, USA: U.S. Department of Agriculture, Forest Service, Pacific Northwest Research Station; 1995. DOI: 10.2737/PNW-GTR-351

[21] Turner MG. Landscape ecology: What is the state of the science? Annual Review of Ecology, Evolution, and Systematics. 2005;**36**:319-344. DOI: 10.1146/annurev.ecolsys.36.102403.120416

[22] Gustafson EJ. Quantifying landscape spatial pattern: What is the state of the art? Ecosystems. 1998;**1**(2):143-156. DOI: 10.1007/s100219900011

[23] Legendre P, Legendre L. Numerical Ecology. 3rd ed. Amsterdam, The Netherlands: Elsevier; 2012. DOI: 10.1016/C2009-0-21590-0

[24] Magurran AE. Measuring Biological Diversity. Oxford, UK: Blackwell Publishing; 2004

[25] Blake GR, Hartge KH. Bulk density. In: Klute A, editor. Methods of Soil Analysis: Part I. Physical and Mineralogical Methods. Madison, WI, USA: American Society of Agronomy; 1986. pp. 363-375

[26] Black CA. Methods of Soil Analysis: Part I. Physical and Mineralogical Properties. Madison, WI, USA: American Society of Agronomy; 1965

[27] Sokal RR, Rohlf FJ. Biometry: The Principles and Practice of Statistics in Biological Research. 3rd ed. New York, NY, USA: W. H. Freeman; 1995

Chapter 4

# Landscape Ecology and Ecosystem Services as Landscape Analysis and Assessment Tools for Ecological Landscape Planning

*Ilda Vagge, Gemma Chiaffarelli, Luigino Pirola, Maddalena Gioia Gibelli and Nicolò Sgalippa*

## Abstract

Landscape ecology, with its models and indicators and ecosystem services, can be the key to interpreting the processes and characteristics of a landscape. An applied methodology is presented in the case study of a peri-urban context (south of the city of Milan, northern Italy), where the urban-rural boundary is often abrupt and unmitigated, a characteristic that significantly increases ecological vulnerability. The multi-scale study, both in space and time, using landscape ecology and ecosystem services, helps to understand the structure of the landscape, its dynamics and the ecological behaviour of the landscape element system. It will also help to identify drivers of vulnerability and resilience, as well as design and planning solutions to rebalance the urban-rural relationship, increase environmental sustainability and resilience to anthropogenic pressures and climate change, and enhance ecosystem services.

**Keywords:** ecological rehabilitation, environmental sustainability, resilient landscape, multi-scale analyses, peri-urban landscape

## 1. Introduction

The rapid and widespread changes in the landscape that have taken place in recent years are linked to changes in land use for social, political and economic reasons as well as to climate change, which cause, among other effects, the loss of biodiversity, the extinction of species and ecosystems, and the degradation of habitats with direct consequences for the possibility of life on the planet. In addition, the conservation of nature, ecosystems and biodiversity must increasingly be reconciled with the growth of the world's population, the depletion of natural resources and the need for food and raw materials. All this makes ecological landscape planning and environmental-ecological design all the more necessary, where *Landscape Ecology* becomes the discipline that can help us better than others to understand the phenomena at play, assess the state of health of a territory and find the most appropriate design solutions to face the environmental challenges that await us. There are many works that explain the power of landscape ecology in ecological landscape planning and design, from the more "classical" and general ones (citing, e.g. [1–7]) to the more recent and technical ones well summarised and discussed in the review works of Milovanovic et al. [8] and Hersperger et al. [9]. In all these works, the *landscape* is interpreted as a *mosaic of ecosystems* (natural, semi-natural and artificial) with its own *structure*, that is, the spatial relationship between patches, corridors and the matrix, *functions*, determined by ecological processes (e.g. flows of energy, materials, animals and plants across the landscape), and *changes*, determined by the interaction of structure and functions over time [3]. The concepts and theories of landscape ecology fit well with the concept of *Ecosystem Services* [10], understood as services provided by landscapes, as a set of ecosystems, and demanded by humans [11]. The concept of ecosystem services is able to link environmental and ecological aspects with productive and cultural ones, providing a general framework for valuing all aspects of the landscape.

This chapter presents a methodological approach, mainly based on Landscape Ecology and Ecosystem Services, which is useful for ecological landscape design and planning, starting with its analysis, interpretation and evaluation. This methodology is applicable to different landscapes, from urban to agricultural [12–15]. In particular, an application to a peri-urban landscape is presented, which is a transitional area between urban and rural environments. This area is particularly complex and interesting due to the conflict between the dynamics of urbanisation and agricultural production. This conflict has led to the phenomena of disturbance and fragmentation of ecosystems, the genesis of simplified and impoverished natural ecosystems, and the lack of socio-cultural services [16]. This is a challenging example of our proposed methodology, which is capable of functioning at multiple *spatiotemporal scales*, thus enabling analysis at different levels of a *hierarchical system*. Hierarchy theory provides a framework for analysing complex systems. Landscape Ecology recognises that landscape moves across many spatial and temporal scales, and while landscape can be very heterogeneous at one scale, it can be quite homogeneous at another [17–20].

## 2. Applied methodology

The methodology used is multidisciplinary and multi-scale, and consists of several successive steps, summarised in **Figure 1** and further clarified in the case study (see Section 3).

Once the objective of the work is clear, for example, the ecological planning of a territory or the environmental design of an area, the first step is to define the areas to be analysed at the different scales, using not administrative but geomorphological and hydrographic limits (in the absence of which linear infrastructures such as roads can be used). The perimeter of the area at the extra-local scale is defined as being wide enough to include all the different landscape features of the place. We then proceed with the definition of the local scale area, which coincides with the ecological planning area and includes any site-specific design areas within it. This is followed by analyses that include both bibliographic and cartographic data as well as site-specific surveys. For the analyses at the *extra-local scale*, the appropriate indices are used, depending on the objective of the work and the territorial context, as summarised in

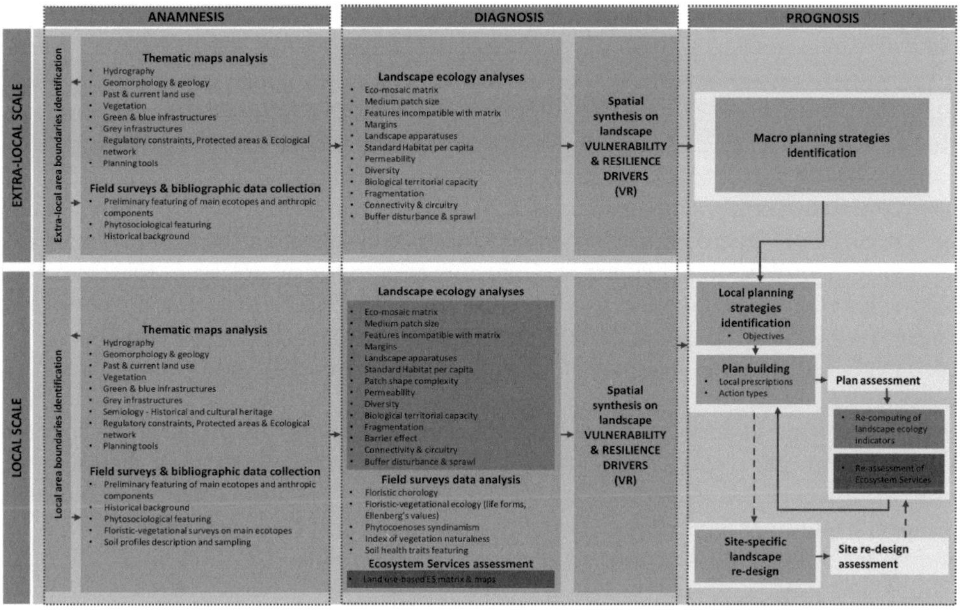


**Figure 1.**
*The proposed multi-scale science-to-practice methodology based on landscape ecology and landscape bionomy approaches and land use-based ecosystem services assessment, to orient strategic landscape ecological planning and to assess plan impacts.*

**Table A1** and **Figure 1**. At the end of the analysis, the landscape drivers of vulnerability and resilience are identified. It is clarified that *vulnerability* means the quality or state of being exposed to the possibility of attack, degradation or damage, with the consequent possibility of modification or loss of landscape characteristics; it is therefore a matter of assessing the sensitivity of the landscape to human use [31–33]. Conversely, *resilience* refers to the ability to maintain ecological, landscape and socio-cultural processes and functions in the presence of anthropogenic pressures and environmental changes [9, 34, 35]. Next, planning macro-strategies to strengthen landscape resilience and mitigate or reduce landscape vulnerability are identified. Then comes the analysis at the *local scale* through the calculation of appropriate indicators and the assessment and mapping of ecosystem services.

For the assessment of ecosystem services, the methodology of Burkhard et al. [36] is applied, suitably adapted to each case study [12, 14], taking into account the classification of ecosystem services according to TEEB [37] and the land uses of a 'normal' European landscape, based on Corine Land Cover Level 3 classes [38]. At the end of the analysis, the drivers of vulnerability and resilience will be identified and mapped at the local scale, followed by the construction of the plan, the validity of which will be assessed through the recalculation of indicators and ecosystem services. Specific plan actions (to be implemented in specific projects) are then indicated. These are developed, for example, through the application of nature-based solutions that use natural processes and systems [39, 40] and the implementation of green-blue infrastructure, which forms a network of aquatic and terrestrial systems in urban and rural landscapes and provides socio-economic and environmental benefits in support of ecosystem functions and societal well-being [41].

## 3. Application on a peri-urban case study

We present an application of the methodology to a peri-urban case study located in Milan city's south-eastern edges (Po Plain, northern Italy) (**Figure 2A**). The area belongs to the Vettabbia Valley, a peri-urban landscape with a deep agricultural historical background which underwent significant structural and functional changes during the last century, shifting this area to a mixed productive-industrial-residential district, currently leaching peri-urban ecosystem services of the past [16]. These ongoing transformation processes brought this area to the attention of local stakeholders, citizens and researchers, who committed to a collaborative pathway to rehabilitate its dormant ecological and cultural values through a multi-stakeholder approach based on agroecology principles (Milano PortaVerde living lab [42]). Several research and science-to-practice activities have already focused on the ecological and cultural values embedded in this territory and supported by the already existing, but spotted, local regeneration projects [12, 16, 43–49]. The *liason* between the past and current traits of this landscape is kept by the Vettabbia irrigation ditch, which flows in a Holocene valley with Gleysols, embedded in fluvial and fluvioglacial sediments (middle hydromorphic Wurm alluvial plain), with Cambisols as prevailing soil type (**Figure 3A**) [50].

The area belongs to a temperate continental bioclimate, upper meso-temperate and upper sub-humid [51]. In line with bioclimate and pedological conditions, the vegetational context of the study area belongs to the *Western neutron-acidophilus Po Plain series of higher and lower plain oak-hornbeam forests* (*Carpinion betuli* Isler 1931 alliance), climacic series of the second fluvial terraces of the low alluvial plain [52]. Due to recurrent and extended anthropic disturbance, secondary substitution series are predominant, with degraded synanthropic spontaneous phytocoenoses, with several invasive exotic species often forming monospecific phytocoenoses [16].

Because of the alluvial and anthropic traits of this peri-urban landscape, the main 'barriers' delimiting the landscape functional organisation can mainly be identified in the hydro-geomorphological and infrastructural components, coupled to the main land use traits. Based on this, we identified extra-local (**Figure 2B**) and local (**Figure 2C**) study areas for applying our methodological approach: these double-scale areas can be considered as interacting units, able to represent the main ecological processes driving their dynamic state.

### 3.1 Extra-local scale

#### *3.1.1 Anamnesis*

The diachronic analysis of land use change (**Figure 3C,D**) shows significant changes occurred since the 50s: there is a great expansion of residential, industrial and other anthropic land uses entailing soil sealing, to the detriment of the peri-urban agricultural matrix, which is nowadays fragmented by several grey linear infrastructures. Linear landscape features were also significantly depleted, whereas there is an increase in wooded surfaces and parks (which still remain a minor component of the landscape). The extra-local area is significantly shaped by blue infrastructures (three main NW-SE axes of main channels and rivers), whereas in-plan green infrastructures flowing out from the city (regional ecological network) currently appear underequipped and highly discontinuous (**Figure 3B**). The presence of a regional

*DOI: http://dx.doi.org/10.5772/intechopen.1006355*

**Figure 2.**
*Localisation of the study area among the Milan city south-eastern edges (northern Italy) (A) and boundaries of the extra-local (B) and local (C) scales of analysis. Some significant landscapes of the study area: the urban fringe, contact with uncultivated areas (D), degraded areas (E), agricultural fields (F); conventional agricultural areas (G, I), agroforestry (H), public parks (L), in the background the Abbey of Chiaravalle (M), historic farmsteads (N), the irrigation network (O) with degraded vegetation (P, Q). [Photos by Vagge I].*

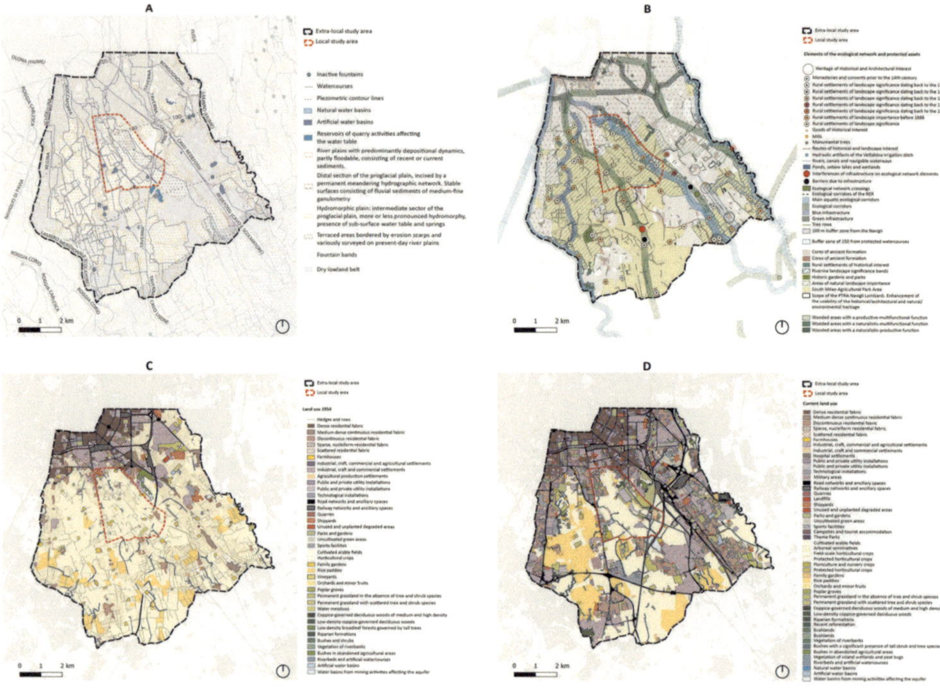


**Figure 3.**
*Anamnesis – thematic maps on the diachronic evolution and current components of the extra-local study area: hydrography and geomorphology (A); elements of the ecological network and protected areas (B); past land use (1954) (C); current land use (D).*

agricultural park distinguishes the remnant agricultural matrix, where several rural heritage buildings persist. In-force planning tools focus on the development of radial green and blue infrastructures (existing parks interconnection, Vettabbia Valley system promotion) and diffused urban ecological regeneration strategies, stressing cultural and citizen accessibility themes within the last urban belt.

### *3.1.2 Diagnosis*

The quantitative investigation of the landscape eco-mosaic composition (**Table 1**, **Figure 4**) highlights the shift from a highly stable peri-urban agricultural matrix in the 50s (83%) to a current unstable and vulnerable agricultural matrix (40%, significantly lower than the 60% threshold [4]); this entails a risk of loss of the agricultural traits of the landscape. Current natural components are strongly limited (7% of total surface). The agricultural patches are highly over-sized (mean patch size: 6.30 ha), impacting on biodiversity-support functions (**Figure 4A**) [53]. The impact of urbanised components incompatible with the agricultural matrix is significant (21.24%). The Standard Habitat index classifies this area as a high-density urban landscape (HS between 80 and 160 $m_2$/inhabitant), lacking in protective functions (HS-protective25 $m_2$/inhabitant), which are secondary to the productive, subsidiary and residential ones (**Table 1**). The overall mean biological territorial capacity (BTC) is low (0.86 Mcal/ha/yr), lower than the regional mean (2.0 Mcal/ha/yr), and this entails high vulnerability of the extra-local landscape (**Table 1**). The Human Habitat (HH) BTC is below the critical threshold (0.79 vs. 1 Mcal/ha/yr); the Natural Habitat (HN) BTC could be significantly raised (current value: 2.19 Mcal/ha/yr). The BTC

| | Eco-mosaic composition | | | | | Habitat | Biological territorial capacity | | Permeability | Diversity | | | Sprawl | |
|---|---|---|---|---|---|---|---|---|---|---|---|---|---|---|
| | Surface | Matrix 50's | Matrix current | Mean patch size | Incompatibles with matrix | Standard habitat (HS) | Mean BTC tot | BTC HN/ BTC tot | Index of draining surface | DIV | DIV/DIV MAX | % DIV | HIGH Sprawl risk areas | Sprawl index |
| U.o.M. | ha | % | % | ha | % | $m_2$/in. | Mcal/ ha/yr | % | % | | % | % | % | $m_2/m_2$ |
| Total | 8492 | | | | | | 0.86 | 14.24 | 62.51 | 2.70 | 68.9 | | 66.99 | 1.32 |
| Agricultural | | 83 | 40 | 6.30 | | | | | | 0.80 | 31.2 | 29.7 | | |
| Natural | | 2 | 7 | 1.55 | | | | | | 0.33 | 12.4 | 12.1 | | |
| Anthropic | | 15 | 53 | 6.29 | | | | | | 1.57 | 43.8 | 58.2 | | |
| Industrial | | | | | 1.4 | | | | | | | | | |
| Urbanised | | | | | 21.2 | | | | | | | | | |
| HN | | | | | | | 2.19 | | | | | | | |
| HH | | | | | | 239.4 | 0.79 | | | | | | | |
| HH – protective | | | | | | 25.0 | | | | | | | | |
| HH – productive | | | | | | 91.7 | | | | | | | | |
| HH – residential | | | | | | 44.8 | | | | | | | | |
| HH – subsidiary | | | | | | 77.9 | | | | | | | | |

*See Appendix A, **Table A1**, for details on the applied indices.*

**Table 1.**
*Results on landscape ecology indicators were calculated for the extra-local study area.*

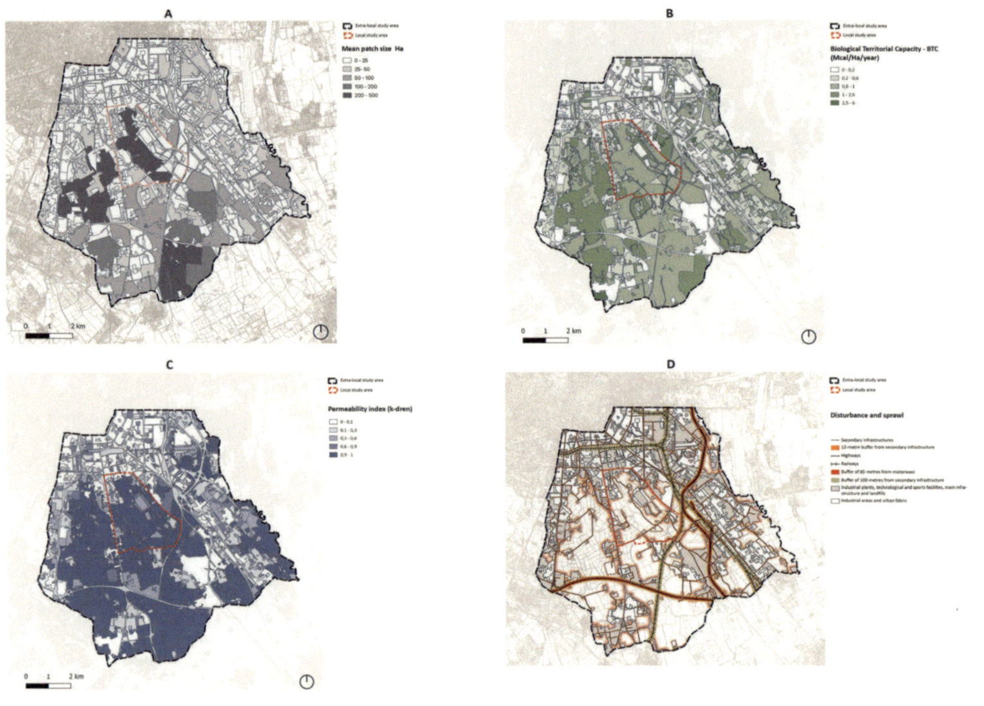


**Figure 4.**
*Diagnosis – maps representing the spatial configuration of some of the indicators computed on the extra-local study area: mean patch size (A); biological territorial capacity (B); permeability (C); disturbance and sprawl (D).*

map (**Figure 4B**) highlights the dichotomy between the urban and the agricultural features, the important role played by the green areas interspersed within the urban matrix, and the absence of significant high BTC areas.

The draining surface index is below 65% (**Table 1**), and this also reflects high vulnerability levels from the hydrological point of view. This vulnerability is significantly increased by the spatial configuration of draining surfaces (**Figure 4C**): they are bundled in large impermeable clusters. The landscape shows positive diversity values (2.70; **Table 1**), but this should be negatively interpreted as its diversity is

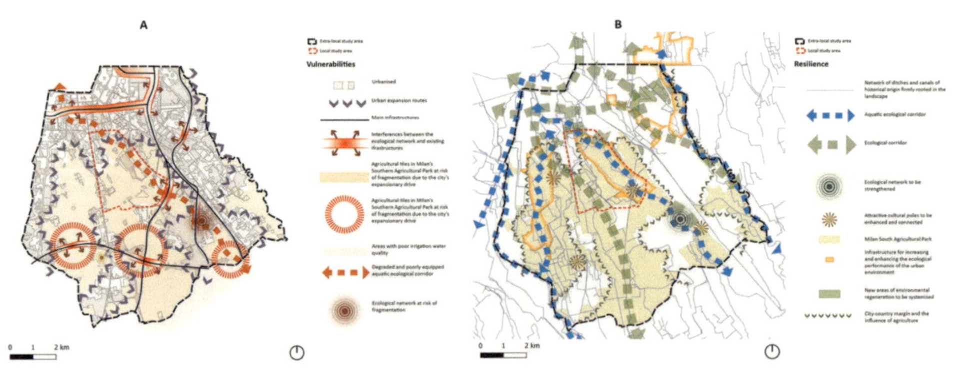


**Figure 5.**
*Diagnosis – Maps synthetising the vulnerability (A) and resilience (B) drivers of the extra-local study area.*

mainly sustained by anthropic components diversification (1.57; 58.2% of total diversity), whereas natural and agricultural components show low diversity values [respectively: 0.33 (29.7% of total diversity); 0.80 (12.1%)]. Urban sprawl impact is consistent: 70% of surface is under high sprawl risk (**Table 1**, **Figure 4D**).

The VR maps synthetise the vulnerability and resilience drivers of the extra-local study, which are brought out by the anamnesis and diagnosis analyses (**Figure 5**).

### 3.2 Local scale

#### *3.2.1 Anamnesis*

Anamnesis phase at the local scale focuses on targeted specific themes relevant to the understanding of local landscape peculiarities. Concerning land use, the local study area is mainly composed of crop fields, with some relevant natural areas of recent insertion (Vettabbia Park, born around the city's water treatment plant) (**Figure 6B**). The semiology map (**Figure 6A**) highlights the historical protagonism and persistence of the water system in the landscape (irrigation ditches network), which drives the landscape texture (**Figure 6B**); also, it shows the persistence of several historical rural buildings, mostly in abandon. All agricultural areas belong to the Regional South Milan Agricultural Park. The Vettabbia irrigation ditch constitutes the main blue infrastructure (NW-SE) to be enhanced (**Figure 6C**). Another ecological corridor is planned on the western belt, but it is currently lacking in structure and continuity.

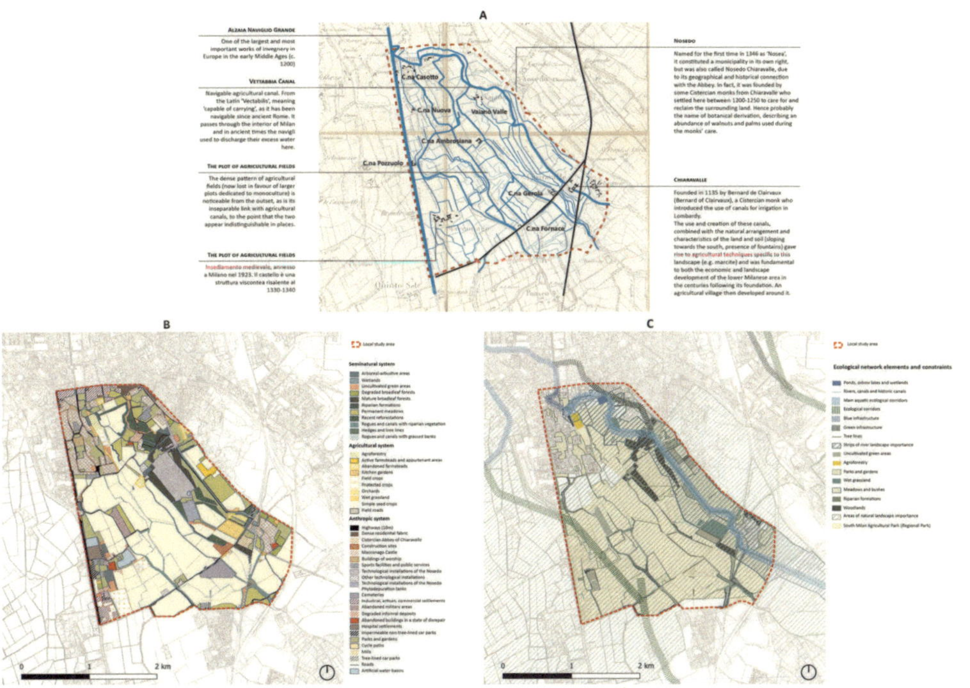


**Figure 6.**
*Anamnesis – thematic maps on the diachronic evolution and current components of the local scale study area: semiology (A); current land use (B); ecological network elements and constraints (C).*

| | Eco-mosaic composition | | | | Biological territorial capacity | | Permeability | Diversity | | | Sprawl | | Fragmentation | | Connectivity | | |
|---|---|---|---|---|---|---|---|---|---|---|---|---|---|---|---|---|---|
| | Surface | Matrix current | Mean patch size | Incompatibles with matrix | Mean BTC | BTC HN/Mean BTC | Index of draining surface | DIV | DIV/DIV MAX | % DIV | HIGH Sprawl risk areas | Sprawl index | Street density | Street fragmentation coefficient | Links/ nodes | Connectivity index | Circuitry index |
| U.o.M. | ha | % | ha | % | Mcal/ ha/yr | % | % | | % | % | % | $m_2/m_2$ | $m/m_2$ | $m_2/m$ | | | |
| Total | 805 | | 1.80 | | 1.15 | 35.8 | 83.3 | 2.51 | | | 70 | 3.31 | 0.004 | 245 | 0.7 | 0.24 | −0.15 |
| Agricultural | | 47 | 2.44 | | | | | 0.60 | 27.2 | 23.8 | | | | | | | |
| Natural | | 20 | 0.84 | | | | | 0.72 | 30.0 | 28.7 | | | | | | | |
| Anthropic | | 33 | 1.98 | | | | | 1.19 | 36.1 | 47.4 | | | | | | | |
| Urbanised | | | | 56 | | | | | | | | | | | | | |
| Industrial | | | | 12.4 | | | | | | | | | | | | | |
| Infrastructures | | | | 6.8 | | | | | | | | | | | | | |
| HN | | | | | 2.98 | | | | | | | | | | | | |
| HH | | | | | 0.85 | | | | | | | | | | | | |

*See Appendix A, **Table A1**, for details on the applied indices.*

**Table 2.**
*Results on landscape ecology indicators calculated on the local scale study area.*

### *3.2.2 Diagnosis*

Landscape ecology indices application at local scale (**Table 2**) highlighted the presence of an agricultural matrix, which is still unstable (56%) but slightly more consistent if compared to the extra-local scale. Natural components are well represented (20.1%), but their spatial configuration might better be spread across the whole local landscape (**Figure 6B**). Main disturbance and threats to the agricultural matrix come from the urbanised areas (56%), industrial areas (12.4%) and, secondarily, from infrastructures (6.8%), as components incompatibles with the matrix from an ecological point of view (**Table 2**). Overall, BTC is higher than the extra-local scale (1.15 Mcal/ha/yr), even though most patches have low BTC values (**Figure 7B**), reflecting the general oversimplification of the agricultural landscape, especially in the central and southern areas. This entails habitat loss and impairment of regulation processes, and consequently, higher landscape dependence on external energetic inputs to preserve its current functions. This results in low self-sustaining capacity of the landscape and significant vulnerability. Connectivity functions are mostly centred on N-S axes, whereas there are no E-W connections (**Figure 7A**); quantitative indicators testify for insufficient connectivity and circuitry (**Table 2**): links/nodes ratio is low (<1), as well as connectivity index (0.24); circuitry index is even negative. The draining surface equals 83%, representing a medium level of vulnerability; the intensive soil sealing in the northern urbanised belt entails significant hydrological unbalances (**Figure 7C**). Landscape diversity is quite good (2.51), but it is mainly supported by anthropic components (47.4% of total diversity) as for the extra-local scale.

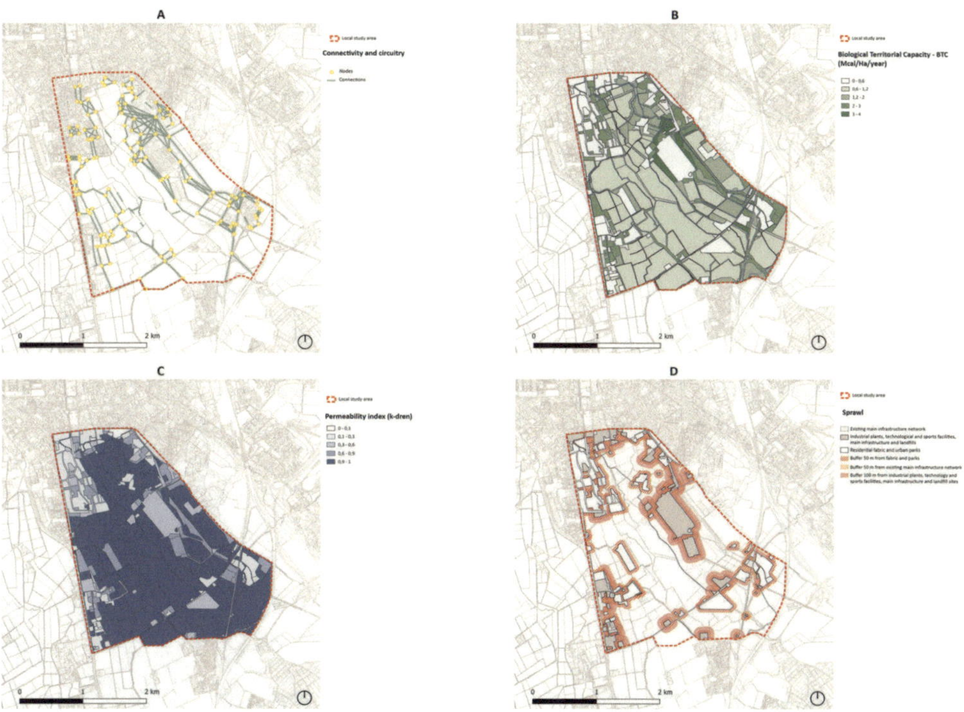


**Figure 7.**
*Diagnosis – maps representing the spatial configuration of some of the indicators computed on the local scale study area: connectivity and circuitry graph (A); biological territorial capacity (BTC); permeability (C); disturbance and sprawl (D).*

Natural and agricultural components are too homogeneous (0.72 and 0.60 diversity values). The sprawl effect is consistent (70% of high sprawl risk areas), especially if coupled to the already fragile agricultural matrix and to fragmentation impacts (fragmentation is quite relevant: 246 $m^2$/linear m) (**Figure 7D**).

The ecosystem services assessment at the local scale, made through the ES matrix (**Figure A1**) and ES maps (**Figure 8**), highlights a general predominance of low support (ES_SUPP) and regulating ecosystem services (ES_REG) delivering values, with some higher hotspots along the eastern ecological axis of the Vettabbia ditch. Urbanised areas and the central-southern agricultural areas lack life support and regulating functions, which would be strongly needed to counterbalance the local and extra-local peri-urban vulnerabilities. Provisioning ES (ES_PROV) are well represented and might better interact with the surrounding urban system and benefit from crop diversification. Cultural ES (ES_CULT) is supported by some localised hotspots linked to the scattered natural components, spiritual and cultural heritage and local experiences on peri-urban regeneration projects. They might better be interconnected and promoted.

The VR maps synthetise the vulnerability and resilience drivers of the local study area in its current state, as brought out by the anamnesis and diagnosis analyses (**Figure 9**).

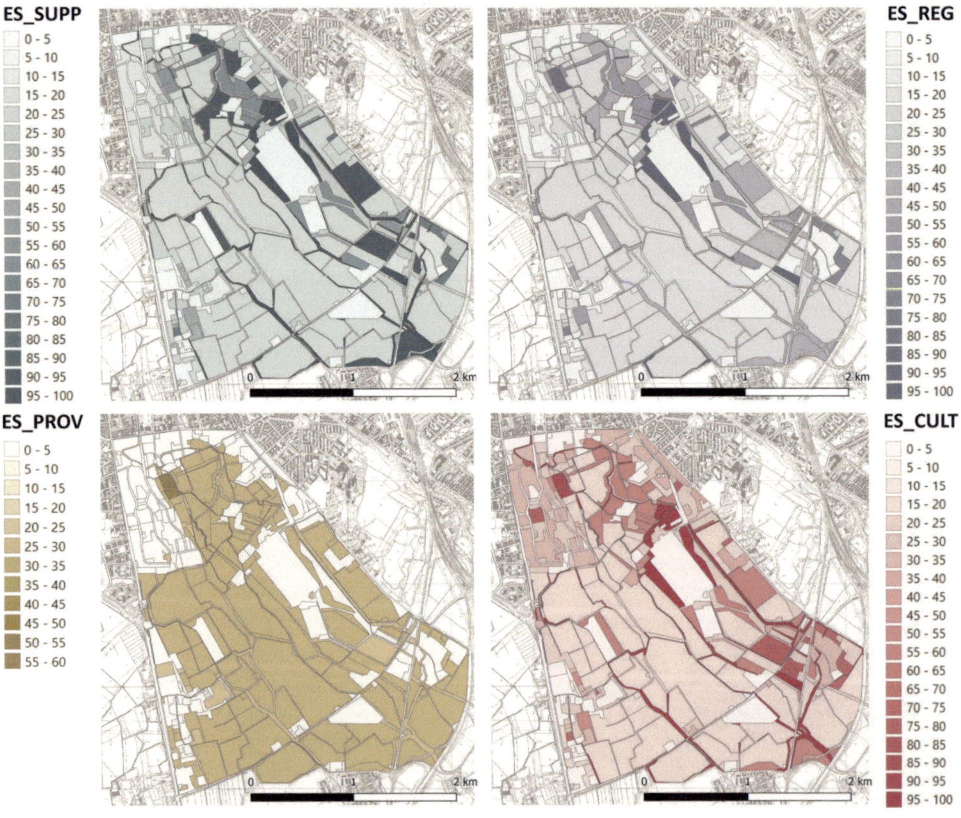


**Figure 8.**
*Diagnosis – maps representing the ecosystem services (ES) delivering capacity of the local scale study area in its current state. From top-left to bottom-right: Support ES (ES_SUPP), Regulating ES (ES_REG), Provisioning ES (ES_PROV), Cultural ES (ES_CULT).*

**Figure 9.**
*Diagnosis – Maps synthetising the vulnerability (A) and resilience (B) drivers of the local study area.*

### *3.2.3 Prognosis*

The analyses at the extra-local and local scales and, specifically, the interpretation of the synthesis on VR maps, allowed us to identify some context-specific local planning strategies to reach focused objectives (**Figure 10A**). The identified strategies aim to solve, or at least balance, the current vulnerabilities of this peri-urban landscape, and to strengthen the resilience drivers capacity to bring this fragile historical landscape to a better ecological balance and to a higher capacity to mitigate and compensate the urban system impacts (extra-local highlighted vulnerabilities). Specifically, life support and regulating functions (and services) could be supported by:

- enhancing the consistency, continuity and ecological quality of the two main NW-SE ecological corridors (requalification of abandoned, degraded green areas; insertion of new linear and areal landscape features; ecological requalification of irrigation ditches), also promoting W-E ecological and cultural interconnections through vegetated paths and the insertion of a new linear park (abandoned railway turned into a greenway);

- mitigating the abrupt urban-rural boundary through buffering belts (ecological re-equipment);

- perforating the urban sealed mineral surfaces through interspersed sustainable drainage systems and nature-based solutions, promoting the sponge effect and stepping stone effect;

- reducing the impact of oversized and over-simplified agricultural areas through new ecological infrastructures, field size reduction and promotion of polyculture.

Provisioning and cultural functions (and services) could be supported by:

- promoting sustainable agricultural practices (organic and regenerative farming practices, crop diversification, agroforestry) through a multi-stakeholder approach inspired by agroecology principles, enabling cultural and knowledge exchanges between local actors and citizens, also thanks to the reactivation of cultural and innovation hubs across the local landscape and their interconnection through green and blue soft mobility paths;

- reallocating the abandoned rural buildings to local cultural hubs and rehabilitating local cultural heritage are new driving forces of this peri-urban landscape identity.

These strategies were then declined at the local scale into a detailed strategic plan, where local prescriptions and targeted action types are integrated within the current landscape texture (**Figure 10B**).

We then assessed the identified plan contributions by reapplying the most relevant landscape ecology indicators: **Table 3** reports the indicator values on the plan scenario and the percentage gaps compared to the current state (difference between plan and current values, divided by current value). Natural components rise by +2.6% and rise

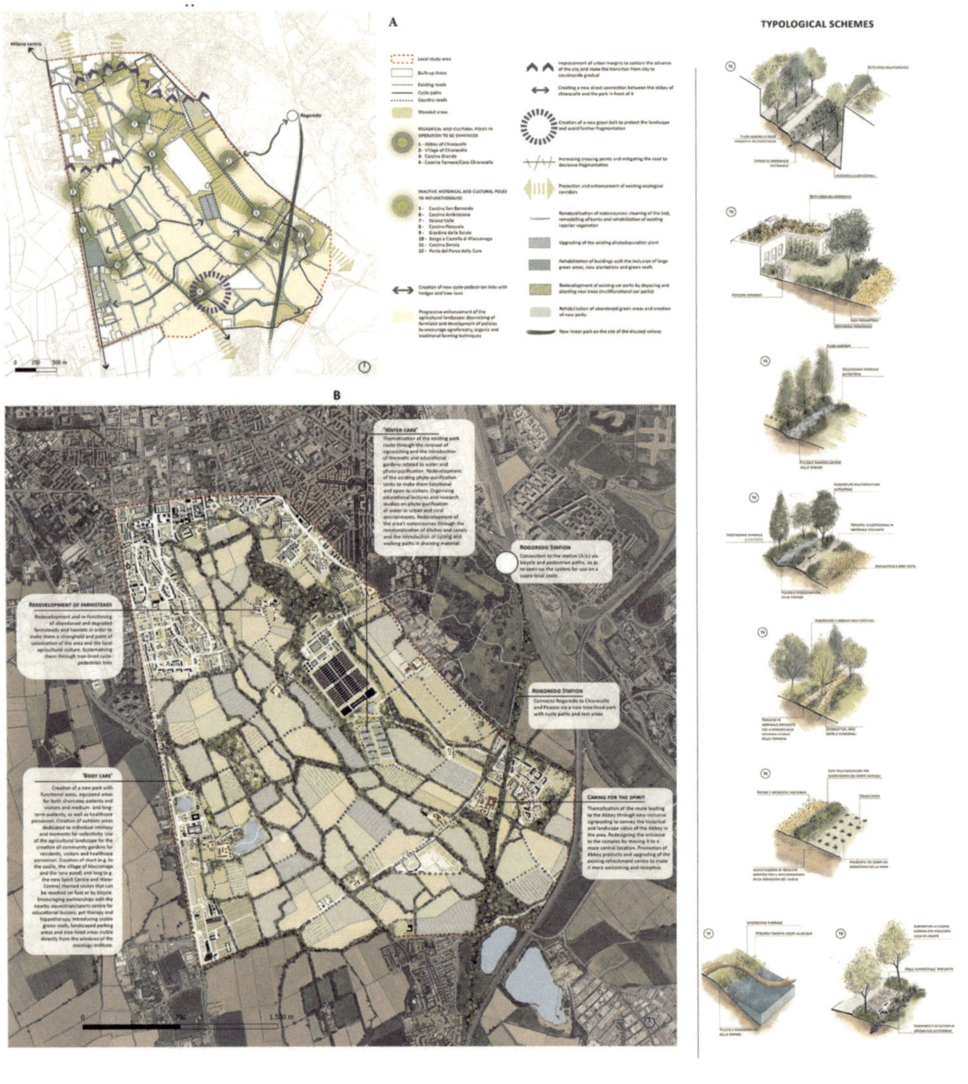


**Figure 10.**
*The identified planning macro-strategies to solve current vulnerabilities of the local scale study area (A) and the detailed plan implementing these strategies (B); on the right side, typological schemes of the selected intervention types are reported. (Graphic design in collaboration with Troisi M. and Boschini L.)*

| | Eco-mosaic composition | | | | | Biological territorial capacity | | | | Permeability | | Diversity | | | | | |
|---|---|---|---|---|---|---|---|---|---|---|---|---|---|---|---|---|---|
| | **Surface** | **Matrix** | | **Mean patch size** | | **Mean BTC** | | **BTC HN/ mean BTC** | | **Index of draining surface** | | **DIV** | | **DIV/DIV MAX** | | **% DIV** | |
| **U.o.M.** | **ha** | **%** | ***gap*** | **ha** | ***gap*** | **Mcal/ha/yr** | ***gap*** | **%** | ***gap*** | **%** | ***gap*** | | ***gap*** | **%** | ***gap*** | **%** | ***gap*** |
| Total | 805 | | | 1.34 | *−25.6* | 1.37 | *+19.1* | 47.5 | *+32.7* | 86 | *+3.2* | 2.80 | *+11.7* | 80.8 | *+24.1* | | |
| Agricultural | | 47.3 | *−0.1* | 1.84 | *−24.6* | | | | | | | 1.01 | *+69.0* | 52.0 | *+91.2* | 36.1 | *+51.5* |
| Natural | | 20.6 | *+2.6* | 1.24 | *+48.2* | | | | | | | 0.73 | *+1.4* | 30.5 | *+1.6* | 26.1 | *−9.1* |
| Anthropic | | 32.1 | *−1.5* | 1.16 | *−41.3* | | | | | | | 1.06 | *−10.8* | 40.1 | *+11.1* | 37.8 | *−20.4* |
| HN | | | | | | 3.26 | *+9.4* | | | | | | | | | | |
| HH | | | | | | 0.9 | *+5.9* | | | | | | | | | | |

*Percentage gaps (compared to current state values) are reported for each applied index. See Appendix A, **Table A1**, for details on the applied indices.*

**Table 3.**
*Results on landscape ecology indicators calculated on the local scale study area, as reconfigured through the identified strategic plan (**Figure 10B**).*

in mean patch size (1.24 ha; +48.2%). Agricultural mean patch size positively diminishes (−24.6%). The overall mean BTC rises to 1.37 Mcal/ha/yr. (+19.1%), with higher values for the HN-BTC, which reflects the increase in natural patches of higher ecological quality (requalification of degraded woodlands and spontaneous tress-shrub phytocoenoses). Draining surface rises +3.2% thanks to the implementation of sustainable urban drainage systems (SUDS) among the urban settlements. Landscape diversity rises to 2.80: it lowers for the anthropic subsystems (which still represent 37.8% of total diversity) while it rises for the agricultural (+69%; 36.1% of total diversity) and the natural areas (+1.4%; 26.1% of total diversity), thanks to crop diversification, agroforestry implementation and landscape features insertion.

We reapplied the ES matrix to the strategic plan updated eco-mosaic (**Figure A2**). The ES maps (**Figure 11**) show a general increase in life support (ES_SUPP) and regulating ecosystem services (ES_REG): the Vettabbia ditch ecological corridor is significantly strengthened, and also other biodiversity source areas are inserted in western areas and among the urbanised matrix. Regulating functions, which were the most needed according to the multi-scale landscape ecological features, show a general rise, with spotted strategic contributions among the agricultural matrix and at the rural-urban interface. Provisioning ES (ES_PROV) increased thanks to crop diversification and agroforestry implementation. Cultural ES (ES_CULT) also show a

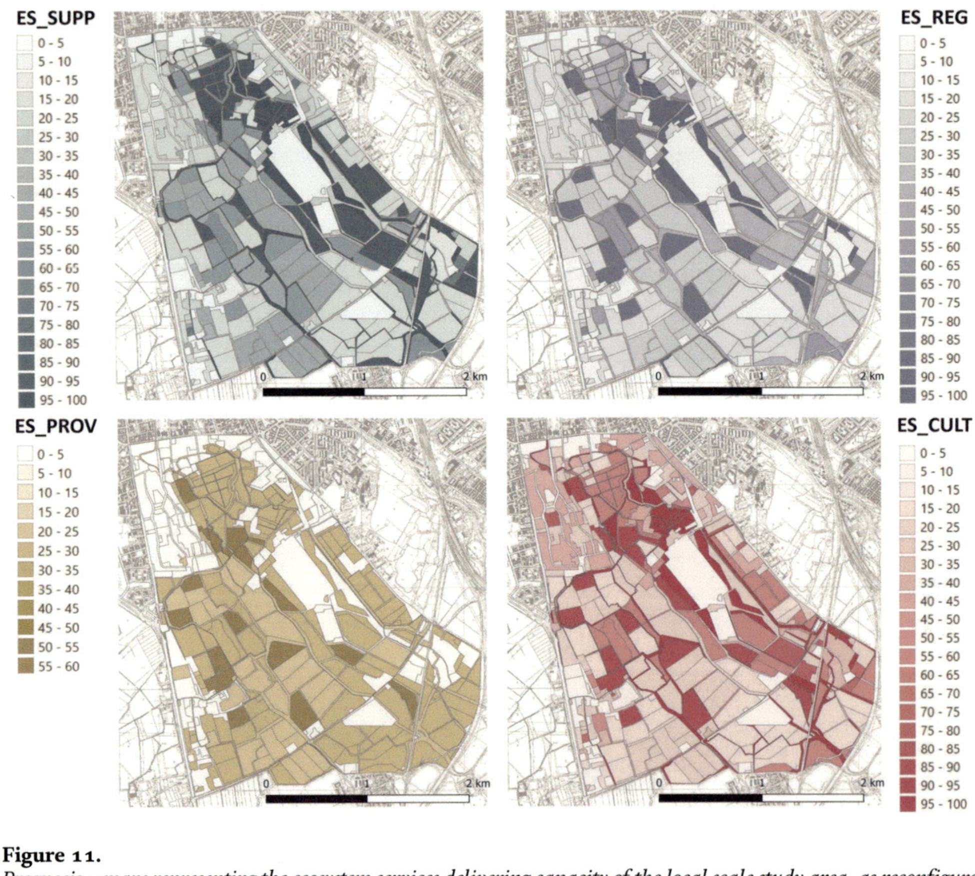


**Figure 11.**
*Prognosis – maps representing the ecosystem services delivering capacity of the local scale study area, as reconfigured through the identified strategic plan (**Figure 10**). From top-left to bottom-right: Support ES (ES_SUPP), Regulating ES (ES_REG), Provisioning ES (ES_PROV), Cultural ES (ES_CULT).*

significant increase: cultural functions are more interspersed across the whole local landscape and better interconnected, which might help in the process of recovery of a landscape identity.

## 4. Conclusions

Ecological landscape planning represents a scientific approach to integrating ecological principles into landscape planning and design. It entails the examination of the interrelationships between living organisms, their habitats and the physical environment, with the objective of promoting biodiversity, conserving natural resources and enhancing ecosystem services, thereby creating sustainable and functional landscapes. This approach acknowledges the necessity of preserving ecological processes and functions, such as nutrient cycling, water filtration and pollination, to ensure the sustenance of healthy ecosystems and human well-being. As previously stated in the introductory chapter, this vision has its roots in the work of Ian McHarg [1], who developed a design approach that takes into account ecological principles. This approach focuses on the integration of natural systems and human development to create sustainable and functional landscapes. These concepts were subsequently taken up, developed, and theorised in numerous works, and landscape ecology became the discipline of reference for ecological landscape design. For example, Forman and Godron present the fundamental concepts of landscape ecology in their book Landscape Ecology [3], elucidating the significance of spatial patterns, connectivity and ecological processes in landscape planning and management. One may consider the work of Wu [54], which provides a comprehensive overview of landscape ecology theory and its application in landscape planning. This work emphasises the importance of understanding spatial patterns, ecological processes and human impacts on landscapes. The key principles of ecological landscape planning are as follows:

1. The concept of *connectivity* is of paramount importance in ecological landscape planning. It is essential to design landscapes that facilitate the movement of species and the flow of ecological processes between different habitats. This can be accomplished through the establishment of wildlife corridors, greenways and stepping stones.

2. The utilisation of *native flora*: The use of native plants is fundamental to the promotion of biodiversity and the creation of habitats for wildlife. Furthermore, they necessitate less maintenance and water resources than non-native plants.

3. The *conservation and restoration of habit*ats: It is essential to identify and protect key habitats that support biodiversity and ecosystem services. This may entail the conservation of existing natural areas, the restoration of degraded ecosystems and the creation of new habitats.

4. The concept of *sustainable land use* encompasses a range of practices aimed at ensuring the long-term viability of natural resources and the ecosystems they support. It is recommended that land use practices be promoted which have the effect of minimising negative impacts on the environment. Such practices might include measures to reduce habitat fragmentation, to minimise pollution and to optimise resource use.

5. The *conservation of water resources* is of great importance. Ecological landscape planning projects integrate water harvesting and conservation systems, including the creation of artificial lakes, ponds and sustainable irrigation systems. This has the effect of reducing the use of fresh water and mitigating the risk of flooding.

6. The *management of rainwater* is a crucial aspect of ecological landscape planning. The project considers the management of rainwater through the implementation of techniques such as rain gardens, green roofs and permeable paving. Such solutions assist in the reduction of surface water pollution and the recharging of groundwater.

7. The integration of *green-blue infrastructure*: the incorporation of natural elements, such as wetlands, forests and green spaces, into the built environment provides multiple benefits, including stormwater management, air purification and recreational opportunities.

8. The *Stakeholder engagement*: The involvement of local communities, landowners and other stakeholders in the planning process ensures that their needs and perspectives are taken into account. This can facilitate the development of support for ecological landscape planning and increase the likelihood of successful implementation.

The implementation of ecological landscape planning principles enables decision-makers to develop landscapes that are more resilient and sustainable, thereby facilitating the well-being of both humans and the natural environment.

There are many ecological landscape design and planning projects, some of which are presented below, that have been developed in Northern Italy and are particularly effective in applying the above criteria. An interesting example of ecological landscape planning is the LOS_DAMA! (Landscape and Open Space Development in Alpine Metropolitan Areas) project. This EU Interreg V B project from November 2016 to December 2019 focuses on Alpine metropolitan areas. They are surrounded by mountains, areas of high natural value such as heathland and moorland, agricultural land and forests. The often unspectacular and degraded landscapes of peri-urban transition zones are important for people's connection to green spaces. Despite their importance, these areas are often overlooked and undervalued by policy makers at all levels and by citizens. In this project, the metropolitan city of Turin (North Italy) has developed a multidisciplinary work with a multi-scalar approach. The aim is to recognise the value of the benefits that green and blue infrastructure provide to urban and peri-urban areas and their inhabitants. There were three scales of work: the metropolitan scale of Corona Verde, which includes 90 municipalities; the scale of the hydrographic basin of Stura di Lanzo; and the local scale, focusing on three specific sites (in Settimo Torinese, San Mauro and Mappano). The methodological approach is based on the disciplines of landscape ecology and environmental economics. The analyses carried out identified the main vulnerability and resilience factors for the Corona Verde area. Spatial indicators were used to identify Landscape-Environmental Units (LEUs) and the factors that characterise the landscapes of Corona Verde and the LEUs. Based on the analysis, tools were developed to guide landscape-environmental scenarios. Scenarios in which green and blue infrastructure is the main strategy to respond to identified vulnerabilities, increase resilience in the LEUs and create ecological connections. This overall strategy contributes to the enrichment of the environmental landscape mosaic, the enhancement of its natural capital and the provision of ecosystem services, and the improvement of the resilience of the pilot area [55–58]. The Milan Metropolitan Territorial Plan (PTM) was developed with the same criterion, with the aim of improving and strengthening the green networks for water lamination and heat island mitigation for the purpose of climate change adaptation. In this plan, in addition to planning the green network of the Milan metropolitan area, the most appropriate measures (green and blue infrastructures, nature-based solutions) have been identified for the different contexts in which the territory is divided, in order to mitigate and adapt to the effects of climate change. The design was carried out after the mapping and valuation of some ecosystem services [59, 60]. An excellent example of ecological design is the HUPAC intermodal terminal project in Gallarate (Varese, N-Italy), completed in 2005, where infrastructure can "generate nature" through the application of green and blue solutions. Rainwater from the intermodal terminal is collected, directed to underground tanks, lifted by a pumping system, and, from there, conveyed by gravity to two series of basins that form a wetland of about 4.5 ha. In addition, afforestation and reforestation works will be carried out on approximately 1.5 ha, and bypasses for terrestrial fauna will be built under the terminal to reduce the previous fragmentation. Finally, paths and ancillary works (including the observatory/landmark), signage and planting design for the terminal have been prepared. Almost 20 years later, the area has become an oasis of biodiversity and naturalness, limiting the effects of extreme rainfall events [61–63]. A final ecological landscape project worth mentioning is the "Arco Verde" project in the province of Bergamo, in the pre-alpine area, which is of great biogeographical and nature conservation interest but is also threatened by the widespread expansion of the urban and industrial tissue and by road and rail infrastructure. Moreover, the territorial fragmentation is accentuated by the orientation of the primary ecological structure, which is based on the major river corridors with a North-South trend, and for which it was therefore necessary to set up ecological reconnection actions on the West-East axis. A 35 km ecological corridor has been created to link parks and nature reserves that had become isolated due to heavy urbanisation. In addition, work was carried out to rehabilitate streams, create wetlands to encourage amphibian fauna, and facilitate or discourage wildlife from crossing infrastructure [64].

It is clear from the above and from the projects presented as examples that spatial and site-specific analyses are fundamental to effective ecological design and planning. In particular, multidisciplinary and multi-scale analyses with ecological (especially landscape ecology) and economic disciplines and ecosystem services at their core. This chapter presents a methodology that can provide a specific analytical framework and tools for multi-scale ecological landscape analysis and assessment. The methodology can be applied to different landscape types, calibrating the different analytical tools according to the specificities of each landscape. The presented methodology has been applied to a complex case study in a peri-urban context. The results show the potential of this methodology to guide the ecological rebalancing of the landscape in a consistent way and to inform about its ecological and multifunctional implications.

## Acknowledgements

We would like to thank Mariagiusi Troisi and Leonardo Boschini, students on the inter-university (University of Genoa and University of Milan) master's degree course in landscape architecture, for their help in collecting and processing the case study data.

## Author contributions

Conceptualization, I.V., L.P., N.S. and M.G.C.; methodology, I.V., L.P., N.S. and G.C; and M.G.G.; software, N.S. and G.C.; validation, I.V., L.P., N.S. and G.C; formal analysis, I.V., N.S. and G.C.; investigation, I.V. and G.C.; resources, I.V. and G.C.; data curation, I.V., G.C. and N.S.; writing—original draft preparation, I.V. and G.C.; writing—review and editing, I.V., G.C., N.S. and L.P.; visualisation, I.V., G.C., N.S. and L.P.; supervision, I.V., M.G.G. and L.P.

All authors have read and agreed to the published version of the manuscript.

The data presented in this study are available on request from the corresponding author.

## Conflict of interest

The authors declare no conflict of interest.

# A. Appendix

| | Indicator | Scale | Equation | Ref. |
|---|---|---|---|---|
| Eco-mosaic composition | Matrix (MTX)<br>x = [natural; agricultural; anthropic] | E_La<br>La | $MTX_x = \frac{\sum_{i=1}^{n} A_{i,x} \times 100}{A_{tot,x}}$<br>$A_i$ = total area of each land use categories patch<br>$A_{tot}$ = total area | [21] |
| | Medium Patch Size (MPS)<br>x = [natural; agricultural; anthropic] | E_La | $MPS_x = \frac{\sum_{i=1}^{n} A_{i,x}}{N_{i,x}}$<br>$A_i$ = area of each patch<br>$N_i$ = no. of patches | [21] |
| | Features incompatible with matrix (INC) | E_La<br>La | $INC = \frac{\sum_{i=1}^{n} A_i}{A_{MTX}}$<br>$A_i$ = area of each patch incompatible with MTX<br>$A_{MTX}$ = area of MTX | [4] |
| | Margins quality | E_La<br>La | $MRG = \frac{N_{+MRG}}{N_{totMRG}} \times 100$<br>$N_{+MRG}$ = no. of positive margins<br>$N_{totMRG}$ = total no. of margins | [4, 22] |
| Habitat | Human Habitat ($H_H$) | E_La<br>La | $H_H = \sum_{i,y=1}^{n,m} A_{iy} \times HH\%_y$<br>$HH\%_y$ = percentage of human habitat of $y$-land use type<br>$A_{iy}$ = area of each patch belonging to $y$-land use type | [7] |
| | Natural Habitat ($H_N$) | E_La<br>La | $H_N = \sum_{i,y=1}^{n,m} [A_{iy} \times (1 - HH\%_y)]$<br>$HH\%_y$ = percentage of human habitat of $y$-land use type<br>$A_{iy}$ = area of each patch belonging to $y$-land use type | [7] |
| Landscape apparatuses | Apparatuses' ratio (APP)<br>1. $H_H$:<br>z = [protective; productive; residential; subsidiary]<br>2. $H_N$:<br>w = [skeletal; connective; excretory; resilient; stabilisation] | E_La<br>La | $APP_z = \frac{\sum_{i=1}^{n} A_{iz} \times HH\%_z}{A_{tot}}$<br>$APP_w = \frac{\sum_{i=1}^{n} [A_{iw} \times (1 - HH\%_w)]}{A_{tot}}$<br>$HH\%_y$ = percentage of HH of each land use type belonging to $z$ apparatus<br>$HH\%_w$ = percentage of HH of each land use type belonging to $w$ apparatus | [7] |
| | Standard Habitat ($H_S$) | E_La<br>La | $H_S = \frac{\sum_{i=1}^{n} APP_{i,z}}{N_{inhab}}$<br>$APP_{i,z}$ = surface of i-apparatus of the human habitat ($APP_z$)<br>$N_{inhab}$ = n. of inhabitants | [7] |

| | Indicator | Scale | Equation | Ref. |
|---|---|---|---|---|
| Shape complexity | Shape Complexity Index (SI) | La | $SI = \frac{0.282 * P_{tot} * 10}{\sqrt{A_{tot}}}$<br>$P_{tot}$ = total perimeter<br>$A_{tot}$ = total area | [4] |
| Diversity | Diversity (DIV)<br>x = [total area; natural subsystem; agricultural subsystem; anthropic subsystem] | E_La | $DIV_x = -\sum_{i=1}^{n} \frac{A_{ix}}{A_{tot,x}} \times \ln \frac{A_{ix}}{A_{tot,x}}$ | [21, 23] |
| Biological territorial capacity | Mean Biological Territorial Capacity (MEAN BTC)<br>x = [total area; Human Habitat subset area; Natural Habitat subset area] | E_La<br>La | $BTC_x = \frac{\sum_{i=1}^{m} BTC_i \times A_{ix}}{A_{tot,x}}$ | [7, 24, 25] |
| Fragmentation | Street density (DENS) | E_La<br>La | $DENS = \frac{I_{lenght}}{A_{nat}}$<br>$I_{length}$ = grey infrastructures length<br>$A_{nat}$ = natural/semi-natural components area | [26] |
| | Street fragmentation coefficient (FRAG) | E_La<br>La | $DENS = \frac{A_{nat}}{I_{length}}$<br>$I_{length}$ = grey infrastructures length<br>$A_{nat}$ = natural/semi-natural components area | [26] |
| | Barrier effect (BARR) | La | $BARR = \sum W + T + I + C$<br>$W$ = street width [0,1,2]<br>T = street type [1–3]<br>I = traffic intensity [1–3]<br>C = street crossing [1–3] | [4, 26, 27] readapted |
| Connectivity & circuitry -graph analysis- | Links/Nodes ratio (L/N) | E_La<br>La | $L/N = \frac{L}{N}$<br>L = n. of links<br>N = n. of nodes | [7, 28] |
| | Connectivity (CON) | E_La<br>La | $CON = \frac{L}{[3\times(N-2)]}$L = no. of links<br>N = no. of nodes | [29] |
| | Weighted connectivity (WCON) | La | $WCON = \frac{\sum_{i=1}^{5} L_i \times W_i}{[3\times(N-2)]}$Li = no. of links for each Ecological Quality Class (EQCi = [1–5])<br>Wi = EQCi weight:<br>$W_i = \frac{EQC_i}{EQC_{max}}$ | [15] |

| | Indicator | Scale | Equation | Ref. |
|---|---|---|---|---|
| | Circuitry (CIR) | E_La<br>La | $CIR = \frac{(L-N+1)}{[2\times(N-5)]}$ | [29] |
| | Weighted circuitry (WCIR) | La | $WCIR = \frac{[(\sum_{i=1}^{5} L_i \times W_i)-N+1)]}{[2\times(N-5)]}$<br>Li = no. of links for each Ecological Quality Class (EQCi = [1–5])<br>Wi = EQCi weight (as above) | [15] |
| Sprawl & disturbance | High sprawl risk areas (SPRW_A) | E_La<br>La | $SPRW\_A = \frac{\sum_{i=1}^{n} A_{i,sprw}}{A_{tot}}$<br>$A_{sprw}$ = areas of 50 m buffer on urbanised patches<br>$A_{tot}$ = total area | [22] |
| | Sprawl index (SPRW) | | $SPRW\_A = \frac{\sum_{i=1}^{n} A_{i,sprw}}{A_{built}}$<br>$A_{sprw}$ = areas of 50 m buffer on urbanised patches<br>$A_{built}$ = total built area | [22] |
| | Disturbance (DIST) | E_La<br>La | $DIST = \sum_{i=1}^{n} A_{i,dist}$<br>$A_{dist}$ = areas of 30 m/100 m buffer on grey linear infrastructures | [27] |
| Permeability | Index of draining surface (DRN) | E_La<br>La | $DRN = \frac{\sum_{i=1}^{n} A_{ix} \times \mathrm{Kdrn}_i}{A_{tot}}$<br>Kdrn = drainage coefficient | [7, 30] |

*For each index, the following information is provided: the employed acronym, the applied scale of analysis [extra-local (E_La); local (La)], the applied equation and the related references.*

**Table A1.**
*List of the landscape ecology indices included in the methodology.*

| | | SUPPORT | | REGULATING | | | | | | | | | | | PROVISIONING | | | | | | | | | | | CULTURAL | | | | | TOTAL |
|---|---|---|---|---|---|---|---|---|---|---|---|---|---|---|---|---|---|---|---|---|---|---|---|---|---|---|---|---|---|---|---|
| **ECOSYSTEM SERVICES MATRIX**<br>**CURRENT STATE** | Area [ha] | Habitat & biodiversity | SUPPORT | Global climate regulation (C seq & stock) | Local climate regulation | Air quality regulation | Water flow regulation | Water purification | Nutrients regulation | Erosion regulation | Natural hazard regulation | Pollination | Pest & disease control | REGULATING | Food (crops, fruits, vegetables) | Biomass for energy | Fodder | Food (livestock) | Fibre | Timber | Food (fish) | Wild foods & resources | Biochemicals & medicine | Freshwater | PROVISIONING | Recreation & eco-tourism | Landscape aestetichs & inspiration | Knowledge systems | Religious & spiritual experience | CULTURAL | TOTAL |
| **NATURAL** | | | **47.1** | | | | | | | | | | | **44.5** | | | | | | | | | | | **25.4** | | | | | **45.6** | **40.6** |
| Hedgerows and treelines | 1.49 | 2 | 34.3 | 2 | 2 | 2 | 1 | 2 | 2 | 2 | 2 | 2 | 2 | 38.6 | 0 | 1 | 1 | 0 | 1 | 2 | 0 | 4 | 3 | 0 | 24.0 | 2 | 2 | 2 | 1 | 38.6 | **33.9** |
| Uncultivated area | 9.73 | 2 | 30.0 | 2 | 1 | 0 | 1 | 1 | 1 | 2 | 1 | 0 | 1 | 16.0 | 0 | 1 | 2 | 3 | 0 | 0 | 0 | 5 | 1 | 0 | 24.0 | 1 | 2 | 2 | 0 | 23.8 | **23.4** |
| Uncultivated area with spontaneous trees&shrubs ri-colonization | 9.79 | 2 | 34.3 | 2 | 2 | 2 | 1 | 2 | 2 | 2 | 2 | 2 | 2 | 38.6 | 0 | 1 | 1 | 0 | 1 | 3 | 0 | 4 | 3 | 0 | 26.0 | 2 | 2 | 2 | 1 | 38.6 | **34.4** |
| Young wood | 21.94 | 2 | 45.5 | 2 | 2 | 2 | 1 | 2 | 2 | 2 | 2 | 2 | 2 | 40.9 | 0 | 1 | 1 | 0 | 1 | 3 | 0 | 4 | 3 | 0 | 26.0 | 2 | 2 | 2 | 1 | 40.9 | **38.3** |
| Degraded wood | 7.11 | 3 | 57.6 | 3 | 3 | 3 | 2 | 3 | 3 | 3 | 2 | 2 | 2 | 51.8 | 0 | 1 | 1 | 0 | 1 | 3 | 0 | 4 | 3 | 0 | 26.0 | 3 | 3 | 3 | 2 | 51.8 | **46.8** |
| Wood | 19.38 | 4 | 81.8 | 4 | 4 | 4 | 2 | 4 | 4 | 4 | 3 | 3 | 3 | 73.6 | 0 | 1 | 1 | 0 | 1 | 3 | 0 | 4 | 3 | 0 | 26.0 | 4 | 4 | 4 | 2 | 73.6 | **63.8** |
| Riparian woody belt | 1.01 | 2 | 46.1 | 3 | 3 | 3 | 2 | 3 | 3 | 3 | 2 | 2 | 2 | 51.8 | 0 | 1 | 1 | 0 | 1 | 3 | 0 | 4 | 3 | 0 | 26.0 | 3 | 3 | 3 | 2 | 51.8 | **43.9** |
| Water bodies | 6.28 | 0 | 6.9 | 0 | 0 | 0 | 0 | 0 | 0 | 0 | 0 | 0 | 0 | 2.6 | 0 | 1 | 0 | 0 | 0 | 0 | 3 | 3 | 0 | 4 | 22.0 | 0 | 0 | 0 | 0 | 5.2 | **9.2** |
| Irrigation ditches - herbaceous | 13.73 | 1 | 18.7 | 0 | 0 | 0 | 1 | 1 | 1 | 0 | 1 | 0 | 1 | 7.5 | 0 | 2 | 0 | 0 | 0 | 0 | 3 | 3 | 0 | 4 | 24.0 | 1 | 1 | 1 | 0 | 16.3 | **16.6** |
| Irrigation ditches - riparian vegetation | 38.72 | 4 | 80.0 | 0 | 1 | 0 | 3 | 3 | 3 | 0 | 3 | 0 | 3 | 32.0 | 0 | 2 | 0 | 0 | 0 | 0 | 3 | 3 | 0 | 4 | 24.0 | 4 | 4 | 4 | 2 | 70.0 | **51.5** |
| Wetland | 6.66 | 4 | 75.9 | 1 | 2 | 0 | 4 | 2 | 2 | 0 | 2 | 0 | 2 | 28.8 | 0 | 1 | 0 | 0 | 0 | 0 | 3 | 3 | 0 | 4 | 22.0 | 4 | 3 | 3 | 2 | 56.9 | **45.9** |
| **AGRICULTURAL** | | | **35.8** | | | | | | | | | | | **26.8** | | | | | | | | | | | **28.9** | | | | | **27.0** | **29.6** |
| Agroforestry | 2.86 | 2 | 42.9 | 4 | 4 | 4 | 2 | 4 | 4 | 4 | 3 | 3 | 3 | 64.3 | 5 | 4 | 4 | 4 | 1 | 3 | 0 | 5 | 3 | 0 | 58.0 | 4 | 4 | 4 | 2 | 64.3 | **57.4** |
| Crop field | 357.80 | 1 | 12.3 | 1 | 2 | 1 | 1 | 0 | 1 | 0 | 1 | 1 | 1 | 13.5 | 5 | 0 | 1 | 0 | 1 | 0 | 0 | 0 | 1 | 0 | 16.0 | 1 | 1 | 1 | 0 | 12.3 | **13.5** |
| Protected crops | 0.60 | 0 | 7.7 | 0 | 1 | 0 | 0 | 0 | 0 | 0 | 0 | 0 | 1 | 8.5 | 5 | 1 | 2 | 0 | 2 | 0 | 0 | 1 | 2 | 0 | 26.0 | 0 | 0 | 1 | 0 | 7.7 | **12.5** |
| Horticoltural | 4.83 | 2 | 36.9 | 1 | 2 | 1 | 2 | 0 | 1 | 0 | 1 | 1 | 2 | 20.3 | 5 | 1 | 2 | 0 | 2 | 0 | 0 | 1 | 2 | 0 | 26.0 | 1 | 1 | 2 | 0 | 18.5 | **25.4** |
| Permanent grassland | 38.33 | 4 | 74.3 | 3 | 1 | 0 | 1 | 1 | 2 | 3 | 1 | 0 | 1 | 29.7 | 0 | 1 | 5 | 5 | 0 | 0 | 0 | 2 | 0 | 0 | 26.0 | 2 | 3 | 3 | 0 | 44.1 | **43.5** |
| Orchards | 1.23 | 2 | 42.9 | 1 | 1 | 1 | 1 | 1 | 1 | 1 | 1 | 4 | 2 | 32.9 | 5 | 1 | 2 | 0 | 2 | 0 | 0 | 1 | 2 | 0 | 26.0 | 2 | 1 | 1 | 0 | 25.0 | **31.7** |
| **ANTHROPIC** | | | **5.5** | | | | | | | | | | | **3.3** | | | | | | | | | | | **0.8** | | | | | **12.1** | **5.4** |
| Agricultural buildings | 7.52 | 1 | 12.8 | 0 | 0 | 0 | 0 | 0 | 0 | 1 | 0 | 1 | 1 | 3.8 | 1 | 0 | 0 | 0 | 0 | 0 | 0 | 0 | 0 | 0 | 2.0 | 2 | 1 | 1 | 1 | 28.8 | **11.9** |
| Heritage buildings | 5.19 | 1 | 11.2 | 0 | 0 | 0 | 0 | 0 | 0 | 1 | 0 | 1 | 1 | 3.4 | 1 | 0 | 0 | 0 | 0 | 0 | 0 | 0 | 0 | 0 | 2.0 | 3 | 3 | 3 | 3 | 56.0 | **18.1** |
| Industrial | 23.80 | 0 | 0.0 | 0 | 0 | 0 | 0 | 0 | 0 | 0 | 0 | 0 | 0 | 0.0 | 0 | 0 | 0 | 0 | 0 | 0 | 0 | 0 | 0 | 0 | 0.0 | 0 | 0 | 0 | 0 | 0.0 | **0.0** |
| Roads | 23.81 | 0 | 0.0 | 0 | 0 | 0 | 0 | 0 | 0 | 0 | 0 | 0 | 0 | 0.0 | 0 | 0 | 0 | 0 | 0 | 0 | 0 | 0 | 0 | 0 | 0.0 | 0 | 0 | 0 | 0 | 0.0 | **0.0** |
| Roads - gravel | 2.98 | 0 | 0.0 | 0 | 0 | 0 | 0 | 0 | 0 | 1 | 0 | 0 | 0 | 1.7 | 0 | 0 | 0 | 0 | 0 | 0 | 0 | 0 | 0 | 0 | 0.0 | 0 | 0 | 0 | 0 | 0.0 | **0.4** |
| Bike path | 1.90 | 0 | 0.0 | 0 | 0 | 0 | 0 | 0 | 0 | 0 | 0 | 0 | 0 | 0.0 | 0 | 0 | 0 | 0 | 0 | 0 | 0 | 0 | 0 | 0 | 0.0 | 0 | 0 | 0 | 0 | 0.0 | **0.0** |
| Railway | 2.10 | 0 | 0.0 | 0 | 0 | 0 | 0 | 0 | 0 | 1 | 0 | 0 | 0 | 1.7 | 0 | 0 | 0 | 0 | 0 | 0 | 0 | 0 | 0 | 0 | 0.0 | 0 | 0 | 0 | 0 | 0.0 | **0.4** |
| Dump and construction sites | 14.79 | 0 | 3.2 | 0 | 0 | 0 | 0 | 0 | 0 | 0 | 0 | 0 | 0 | 1.0 | 1 | 0 | 0 | 0 | 0 | 0 | 0 | 0 | 0 | 0 | 2.0 | 0 | 0 | 0 | 0 | 0.0 | **1.5** |
| Other services | 48.92 | 0 | 0.0 | 0 | 0 | 0 | 0 | 0 | 0 | 1 | 0 | 0 | 1 | 3.4 | 0 | 0 | 0 | 0 | 0 | 0 | 0 | 0 | 0 | 0 | 0.0 | 0 | 0 | 0 | 0 | 0.0 | **0.9** |
| Parkings | 6.02 | 0 | 0.0 | 0 | 0 | 0 | 0 | 0 | 0 | 0 | 0 | 0 | 0 | 0.0 | 0 | 0 | 0 | 0 | 0 | 0 | 0 | 0 | 0 | 0 | 0.0 | 0 | 0 | 0 | 0 | 0.0 | **0.0** |
| Parkings with trees | 1.89 | 1 | 12.8 | 0 | 0 | 0 | 0 | 0 | 0 | 1 | 0 | 1 | 1 | 3.8 | 1 | 0 | 0 | 0 | 0 | 0 | 0 | 0 | 0 | 0 | 2.0 | 0 | 0 | 0 | 0 | 0.0 | **4.7** |
| Parks and private green areas | 65.12 | 1 | 18.5 | 1 | 1 | 1 | 1 | 1 | 1 | 1 | 0 | 1 | 1 | 11.7 | 0 | 0 | 0 | 0 | 0 | 0 | 0 | 0 | 0 | 0 | 0.0 | 1 | 1 | 1 | 1 | 16.9 | **11.8** |
| Dense residential buildings | 4.87 | 0 | 0.0 | 0 | 0 | 0 | 0 | 0 | 0 | 1 | 0 | 0 | 1 | 3.4 | 0 | 0 | 0 | 0 | 0 | 0 | 0 | 0 | 0 | 0 | 0.0 | 2 | 1 | 1 | 1 | 25.7 | **7.3** |
| Sparse residential buildings | 54.74 | 1 | 11.2 | 0 | 0 | 0 | 0 | 0 | 0 | 1 | 0 | 1 | 1 | 3.4 | 1 | 0 | 0 | 0 | 0 | 0 | 0 | 0 | 0 | 0 | 2.0 | 2 | 1 | 1 | 1 | 25.2 | **10.4** |

**Figure A1.**
*The ecosystem services matrix applied on the local scale study area (current state eco-mosaic), showing the total surface values for each land use type.*

| | | | | SUPPORT | | REGULATING | | | | | | | | | | | PROVISIONING | | | | | | | | | | | CULTURAL | | | | | TOTAL |
|---|---|---|---|---|---|---|---|---|---|---|---|---|---|---|---|---|---|---|---|---|---|---|---|---|---|---|---|---|---|---|---|---|---|
| **ECOSYSTEM SERVICES MATRIX**<br>**PLAN** | Area plan [ha] | Area current [ha] | Area plan - Area current [ha] | Habitat & biodiversity | SUPPORT | Global climate regulation (C seq & stock) | Local climate regulation | Air quality regulation | Water flow regulation | Water purification | Nutrients regulation | Erosion regulation | Natural hazard regulation | Pollination | Pest & disease control | REGULATING | Food (crops, fruits, vegetables) | Biomass for energy | Fodder | Food (livestock) | Fibre | Timber | Food (fish) | Wild foods & resources | Biochemicals & medicine | Freshwater | PROVISIONING | Recreation & eco-tourism | Landscape aestetichs & inspiration | Knowledge systems | Religious & spiritual experience | CULTURAL | TOTAL |
| **NATURAL** | | | | | **47.1** | | | | | | | | | | | **44.5** | | | | | | | | | | | **25.4** | | | | | **45.6** | **40.6** |
| Hedgerows and treelines | 5.49 | 1.49 | 4.01 | 2 | 34.3 | 2 | 2 | 2 | 1 | 2 | 2 | 2 | 2 | 2 | 2 | 38.6 | 0 | 1 | 1 | 0 | 1 | 2 | 0 | 4 | 3 | 0 | 24.0 | 2 | 2 | 2 | 1 | 38.6 | **33.9** |
| Uncultivated area | 8.87 | 9.73 | -0.86 | 2 | 30.0 | 2 | 1 | 0 | 1 | 1 | 1 | 2 | 1 | 0 | 1 | 16.0 | 0 | 1 | 2 | 3 | 0 | 0 | 0 | 5 | 1 | 0 | 24.0 | 1 | 2 | 2 | 0 | 23.8 | **23.4** |
| Uncultivated area with spontaneous trees&shrubs ri-colonization | 12.31 | 9.79 | 2.52 | 2 | 34.3 | 2 | 2 | 2 | 1 | 2 | 2 | 2 | 2 | 2 | 2 | 38.6 | 0 | 1 | 1 | 0 | 1 | 3 | 0 | 4 | 3 | 0 | 26.0 | 2 | 2 | 2 | 1 | 38.6 | **34.4** |
| Young wood | 20.12 | 21.94 | -1.82 | 2 | 45.5 | 2 | 2 | 2 | 1 | 2 | 2 | 2 | 2 | 2 | 2 | 40.9 | 0 | 1 | 1 | 0 | 1 | 3 | 0 | 4 | 3 | 0 | 26.0 | 2 | 2 | 2 | 1 | 40.9 | **38.3** |
| Degraded wood | 4.14 | 7.11 | -2.97 | 3 | 57.6 | 3 | 3 | 3 | 2 | 3 | 3 | 3 | 2 | 2 | 2 | 51.8 | 0 | 1 | 1 | 0 | 1 | 3 | 0 | 4 | 3 | 0 | 26.0 | 3 | 3 | 3 | 2 | 51.8 | **46.8** |
| Wood | 42.15 | 19.38 | 22.77 | 4 | 81.8 | 4 | 4 | 4 | 2 | 4 | 4 | 4 | 3 | 3 | 3 | 73.6 | 0 | 1 | 1 | 0 | 1 | 3 | 0 | 4 | 3 | 0 | 26.0 | 4 | 4 | 4 | 2 | 73.6 | **63.8** |
| Riparian woody belt | 1.01 | 1.01 | 0.00 | 2 | 46.1 | 3 | 3 | 3 | 2 | 3 | 3 | 3 | 2 | 2 | 2 | 51.8 | 0 | 1 | 1 | 0 | 1 | 3 | 0 | 4 | 3 | 0 | 26.0 | 3 | 3 | 3 | 2 | 51.8 | **43.9** |
| Water bodies | 6.28 | 6.28 | 0.00 | 0 | 6.9 | 0 | 0 | 0 | 0 | 0 | 0 | 0 | 0 | 0 | 0 | 2.6 | 0 | 1 | 0 | 0 | 0 | 0 | 3 | 3 | 0 | 4 | 22.0 | 0 | 0 | 0 | 0 | 5.2 | **9.2** |
| Irrigation ditches - herbaceous | 9.97 | 13.73 | -3.76 | 1 | 18.7 | 0 | 0 | 0 | 1 | 1 | 1 | 0 | 1 | 0 | 1 | 7.5 | 0 | 2 | 0 | 0 | 0 | 0 | 3 | 3 | 0 | 4 | 24.0 | 1 | 1 | 1 | 0 | 16.3 | **16.6** |
| Irrigation ditches - riparian vegetation | 48.66 | 38.72 | 9.94 | 4 | 80.0 | 0 | 1 | 0 | 3 | 3 | 3 | 0 | 3 | 0 | 3 | 32.0 | 0 | 2 | 0 | 0 | 0 | 0 | 3 | 3 | 0 | 4 | 24.0 | 4 | 4 | 4 | 2 | 70.0 | **51.5** |
| Wetland | 6.66 | 6.66 | -0.01 | 4 | 75.9 | 1 | 2 | 0 | 4 | 2 | 2 | 0 | 2 | 0 | 2 | 28.8 | 0 | 1 | 0 | 0 | 0 | 0 | 3 | 3 | 0 | 4 | 22.0 | 4 | 3 | 3 | 2 | 56.9 | **45.9** |
| **AGRICULTURAL** | | | | | **35.8** | | | | | | | | | | | **26.8** | | | | | | | | | | | **28.9** | | | | | **27.0** | **29.6** |
| Agroforestry | 54.90 | 2.86 | 52.04 | 2 | 42.9 | 4 | 4 | 4 | 2 | 4 | 4 | 4 | 3 | 3 | 3 | 64.3 | 5 | 4 | 4 | 4 | 1 | 3 | 0 | 5 | 3 | 0 | 58.0 | 4 | 4 | 4 | 2 | 64.3 | **57.4** |
| Crop field | 151.35 | 357.80 | -206.45 | 1 | 12.3 | 1 | 2 | 1 | 1 | 0 | 1 | 0 | 1 | 1 | 1 | 13.5 | 5 | 0 | 1 | 0 | 1 | 0 | 0 | 0 | 1 | 0 | 16.0 | 1 | 1 | 1 | 0 | 12.3 | **13.5** |
| Crop field in rotation | 98.78 | 0.00 | 98.78 | 2 | 33.8 | 1 | 3 | 1 | 1 | 0 | 1 | 0 | 1 | 1 | 2 | 18.6 | 5 | 1 | 2 | 0 | 2 | 0 | 0 | 1 | 1 | 0 | 24.0 | 1 | 1 | 2 | 0 | 16.9 | **23.3** |
| Protected crops | 0.60 | 0.60 | 0.00 | 0 | 7.7 | 0 | 1 | 0 | 0 | 0 | 0 | 0 | 0 | 0 | 1 | 8.5 | 5 | 1 | 2 | 0 | 2 | 0 | 0 | 1 | 2 | 0 | 26.0 | 0 | 0 | 1 | 0 | 7.7 | **12.5** |
| Horticoltural | 4.83 | 4.83 | 0.00 | 2 | 36.9 | 1 | 2 | 1 | 2 | 0 | 1 | 0 | 1 | 1 | 2 | 20.3 | 5 | 1 | 2 | 0 | 2 | 0 | 0 | 1 | 2 | 0 | 26.0 | 1 | 1 | 2 | 0 | 18.5 | **25.4** |
| Permanent grassland | 68.87 | 38.33 | 30.54 | 4 | 74.3 | 3 | 1 | 0 | 1 | 1 | 2 | 3 | 1 | 0 | 1 | 29.7 | 0 | 1 | 5 | 5 | 0 | 0 | 0 | 2 | 0 | 0 | 26.0 | 2 | 3 | 3 | 0 | 44.1 | **43.5** |
| Orchards | 1.23 | 1.23 | 0.00 | 2 | 42.9 | 1 | 1 | 1 | 1 | 1 | 1 | 1 | 1 | 4 | 2 | 32.9 | 5 | 1 | 2 | 0 | 2 | 0 | 0 | 1 | 2 | 0 | 26.0 | 2 | 1 | 1 | 0 | 25.0 | **31.7** |
| **ANTHROPIC** | | | | | **5.5** | | | | | | | | | | | **3.3** | | | | | | | | | | | **0.8** | | | | | **12.1** | **5.4** |
| Agricultural buildings | 9.85 | 7.52 | 2.32 | 1 | 12.8 | 0 | 0 | 0 | 0 | 0 | 0 | 1 | 0 | 1 | 1 | 3.8 | 1 | 0 | 0 | 0 | 0 | 0 | 0 | 0 | 0 | 0 | 2.0 | 2 | 1 | 1 | 1 | 28.8 | **11.9** |
| Heritage buildings | 5.19 | 5.19 | 0.00 | 1 | 11.2 | 0 | 0 | 0 | 0 | 0 | 0 | 1 | 0 | 1 | 1 | 3.4 | 1 | 0 | 0 | 0 | 0 | 0 | 0 | 0 | 0 | 0 | 2.0 | 3 | 3 | 3 | 3 | 56.0 | **18.1** |
| Industrial | 23.78 | 23.80 | -0.02 | 0 | 0.0 | 0 | 0 | 0 | 0 | 0 | 0 | 0 | 0 | 0 | 0 | 0.0 | 0 | 0 | 0 | 0 | 0 | 0 | 0 | 0 | 0 | 0 | 0.0 | 0 | 0 | 0 | 0 | 0.0 | **0.0** |
| Roads | 23.80 | 23.81 | -0.02 | 0 | 0.0 | 0 | 0 | 0 | 0 | 0 | 0 | 0 | 0 | 0 | 0 | 0.0 | 0 | 0 | 0 | 0 | 0 | 0 | 0 | 0 | 0 | 0 | 0.0 | 0 | 0 | 0 | 0 | 0.0 | **0.0** |
| Roads - gravel | 2.98 | 2.98 | 0.00 | 0 | 0.0 | 0 | 0 | 0 | 0 | 0 | 0 | 1 | 0 | 0 | 0 | 1.7 | 0 | 0 | 0 | 0 | 0 | 0 | 0 | 0 | 0 | 0 | 0.0 | 0 | 0 | 0 | 0 | 0.0 | **0.4** |
| Bike path | 1.90 | 1.90 | 0.00 | 0 | 0.0 | 0 | 0 | 0 | 0 | 0 | 0 | 0 | 0 | 0 | 0 | 0.0 | 0 | 0 | 0 | 0 | 0 | 0 | 0 | 0 | 0 | 0 | 0.0 | 0 | 0 | 0 | 0 | 0.0 | **0.0** |
| Railway | 2.10 | 2.10 | 0.00 | 0 | 0.0 | 0 | 0 | 0 | 0 | 0 | 0 | 1 | 0 | 0 | 0 | 1.7 | 0 | 0 | 0 | 0 | 0 | 0 | 0 | 0 | 0 | 0 | 0.0 | 0 | 0 | 0 | 0 | 0.0 | **0.4** |
| Dump and construction sites | 7.34 | 14.79 | -7.45 | 0 | 3.2 | 0 | 0 | 0 | 0 | 0 | 0 | 0 | 0 | 0 | 0 | 1.0 | 1 | 0 | 0 | 0 | 0 | 0 | 0 | 0 | 0 | 0 | 2.0 | 0 | 0 | 0 | 0 | 0.0 | **1.5** |
| Other services | 48.88 | 48.92 | -0.04 | 0 | 0.0 | 0 | 0 | 0 | 0 | 0 | 0 | 1 | 0 | 0 | 1 | 3.4 | 0 | 0 | 0 | 0 | 0 | 0 | 0 | 0 | 0 | 0 | 0.0 | 0 | 0 | 0 | 0 | 0.0 | **0.9** |
| Parkings with trees | 7.90 | 1.89 | 6.01 | 1 | 12.8 | 0 | 0 | 0 | 0 | 0 | 0 | 1 | 0 | 1 | 1 | 3.8 | 1 | 0 | 0 | 0 | 0 | 0 | 0 | 0 | 0 | 0 | 2.0 | 0 | 0 | 0 | 0 | 0.0 | **4.7** |
| Parks and private green areas | 65.06 | 65.12 | -0.06 | 1 | 18.5 | 1 | 1 | 1 | 1 | 1 | 1 | 1 | 0 | 1 | 1 | 11.7 | 0 | 0 | 0 | 0 | 0 | 0 | 0 | 0 | 0 | 0 | 0.0 | 1 | 1 | 1 | 1 | 16.9 | **11.8** |
| Dense residential buildings | 1.86 | 4.87 | -3.01 | 0 | 0.0 | 0 | 0 | 0 | 0 | 0 | 0 | 1 | 0 | 0 | 1 | 3.4 | 0 | 0 | 0 | 0 | 0 | 0 | 0 | 0 | 0 | 0 | 0.0 | 2 | 1 | 1 | 1 | 25.7 | **7.3** |
| Sparse residential buildings with SUDS | 35.98 | 0.00 | 35.98 | 1 | 12.8 | 1 | 0 | 0 | 2 | 1 | 1 | 1 | 0 | 1 | 1 | 12.8 | 0 | 0 | 0 | 0 | 0 | 0 | 0 | 0 | 0 | 1 | 2.0 | 2 | 1 | 1 | 1 | 28.8 | **14.1** |
| Sparse residential buildings | 21.72 | 54.74 | -33.01 | 1 | 11.2 | 0 | 0 | 0 | 0 | 0 | 0 | 1 | 0 | 1 | 1 | 3.4 | 1 | 0 | 0 | 0 | 0 | 0 | 0 | 0 | 0 | 0 | 2.0 | 2 | 1 | 1 | 1 | 25.2 | **10.4** |

**Figure A2.**
*The ecosystem services matrix applied to the local scale study area, as reconfigured through the identified strategic plan (**Figure 10**). The total surface values for each land use type in the plan scenario are reported and compared to current state values (difference between surfaces: plan minus current state).*

## Author details

Ilda Vagge*, Gemma Chiaffarelli, Luigino Pirola, Maddalena Gioia Gibelli and Nicolò Sgalippa
Department of Agricultural and Environmental Sciences, University of Milan, Via Celoria, Milan, Italy

*Address all correspondence to: ilda.vagge@unimi.it

## References

[1] Mcharg IL. Design with Nature. Garden City, NY: American Museum of Natural History [by] the Natural History Press; 1969

[2] Naveh Z, Lieberman AS. Landscape Ecology: Theory and Application. New York: Springer; 1984

[3] Forman RTT, Godron M. Landscape Ecology. New York, NY, USA: John Wiley and Sons; 1986

[4] Forman RTT. Land Mosaics: The Ecology of Landscapes and Regions. 1st ed. Cambridge, UK: Cambridge University Press; 1995

[5] Farina A. Landscape Ecology in Action. Dordrecht: Springer; 2000. pp. XII, 317

[6] Botequilha-Leitão A, Ahern J. Appling landscape ecological concepts and metrics in sustainable landscape planning. Landscape and Urban Planning. 2002;**59**:65-93. DOI: 10.1016/S0169-2046(02)00005-1

[7] Ingegnoli V. Landscape Bionomics: Biological-Integrated Lanscape Ecology. Milan: Springer; 2015

[8] Milovanović A, Milovanovic RD, Maruna M. Eighty-year review of the evolution of landscape ecology: From a spatial planning perspective. Landscape Ecology. 2020;**35**:1-21. DOI: 10.1007/s10980-020-01102-9

[9] Hersperger AM, Grădinaru SR, Pierri Daunt AB, Imhof CS, Fan P. Landscape ecological concepts in planning: Review of recent developments. Landscape Ecology. 2021;**36**:2329-2345. DOI: 10.1007/s10980-021-01193-y

[10] Costanza R, d'Arge R, de Groot R, Farber S, Grasso M, Hannon B, et al. The value of the world's ecosystem services and natural capital. Nature. 1997;**387**:253-260. DOI: 10.1038/387253a0

[11] Keller R, Backhaus N. Integrating landscape services into policy and practice – A case study from Switzerland. Landscape Research. 2020;**45**:111-122. DOI: 10.1080/01426397.2019.1569218

[12] Dal Borgo AG, Chiaffarelli G, Capocefalo V, Schievano A, Bocchi S, Vagge I. Agroforestry as a driver for the provisioning of Peri-Urban socio-ecological functions: A trans-disciplinary approach. Sustainability. 2023;**15**(14),11020:1-30. DOI: 10.3390/su151411020

[13] Vagge I, Sgalippa N, Chiaffarelli G. The role of agroforestry in solving the agricultural landscapes vulnerabilities in the Po plain district. Community Ecology. 2024. DOI: 10.1007/s42974-024-00203-8

[14] Vagge I, Sgalippa N, Chiaffarelli G. Agricultural landscapes: A pattern-process-design approach to enhance their ecological quality and ecosystem services through agroforestry. Diversity. 2024;**16**(7),431:1-26. DOI: 10.3390/d16070431

[15] Vagge I, Chiaffarelli G. Validating the contribution of nature-based farming solutions (NBFS) to agrobiodiversity values through a multi-scale landscape approach. Agronomy. 2023;**13**(1):233. DOI: 10.3390/agronomy13010233

[16] Chiaffarelli G, Vagge I. Cities vs countryside: An example of a science-based Peri-urban landscape features rehabilitation in Milan (Italy). Urban Forestry & Urban Greening. 2023;**86**: 128002. DOI: 10.1016/j.ufug.2023128002

[17] Allan TFH, Starr TB. Hierarchies: Perspectives for Ecological Complexity.

Chicago: University of Chicago Press; 1982. p. 352

[18] Urban DL, O'Neill RV, Shugart HH Jr. Landscape ecology, a hierarquical perspective can help scientists understand spatial patterns. Bioscience. 1987;**37**:119-127. DOI: 10.2307/1310366

[19] Forman RTT. The ethics of isolation, the spread of disturbance, and landscape ecology. In: Turner MG, editor. Landscape Heterogeneity and Disturbance. New York, NY: Springer New York; 1987. pp. 213-229

[20] Meentemeyer V, Box EO. Scale effects in landscape studies. In: Turner MG, editor. Landscape Heterogeneity and Disturbance. New York, NY: Springer New York; 1987. pp. 15-34

[21] Turner MG, Gardner RH. Landscape Ecology in Theory and Practice, Pattern and Process. New York: Springer Verlag; 2015

[22] Gibelli MG, Sartorio MF, Gibelli G, Lodi M, Santolini R. Landscapes' functions and human health: Incidence of environmental changes. Journal of Mediterranean Ecology. 2007;**8**:27-42

[23] Shannon CE. A mathematical theory of communication. The Bell System Technical Journal. 1948;**27**:379-423. DOI: 10.1002/j.1538-7305.1948.tb01338.x

[24] Ingegnoli V, Giglio E. Proposal of a synthetic indicator to control ecological dynamics at an ecological mosaic scale. Annali di Botanica. 1999;**LVII**:181-190

[25] Ingegnoli V. The study of vegetation for a diagnostical evaluation of agricultural landscapes, some examples fom Lombardy. Annali di Botanica Nuova Serie. 2006;**VI**:111-120

[26] Romano B, Paolinelli G. L'interferenza insediativa nelle strutture ecosistemiche. Roma: Gamgemi editore; 2007

[27] Bellelli L, Bretzel F, Cattaneo E, D'Ambrogi S, Franchi G, Gibelli G, et al. L'inserimento paesaggistico delle infrastrutture stradali: strumenti metodologici e buone pratiche di progetto. In: Bonci L, editor. Ambiente, Paesaggio e Infrastrutture. Vol. 65.5. Roma: Istituto Superiore per la Protezione e la Ricerca Ambientale; 2010

[28] Ingegnoli V, Giglio E. Ecologia del Paesaggio: Manuale per Conservare, Gestire e Pianificare L'ambiente. Napoli: Sistemi editoriali; 2005

[29] Fabbri P. Ecologia del paesaggio per la pianificazione / Pompeo Fabbri. Aracne: Roma; 2005

[30] Ermer K, Hoff R, Mohrmann R. Landschaftsplanung Inder Stadt. Ulmer: Stuttgart, Germany; 1996

[31] Ćurović Ž, Ćurović M, Spalević V, Janic M, Sestras P, Popović SG. Identification and evaluation of landscape as a precondition for planning revitalization and development of Mediterranean rural settlements—Case study: Mrkovi Village, Bay of Kotor, Montenegro. Sustainability. 2019;**11**(7): 2039:2-15. DOI: 10.3390/su11072039

[32] Haber MJ, Hibbert B. The vulnerability landscape. In: Haber MJ, Hibbert B, editors. Asset Attack Vectors: Building Effective Vulnerability Management Strategies to Protect Organizations. Berkeley, CA: Apress; 2018. pp. 5-37

[33] Golobic M, Breskvar ZL. Landscape planning and vulnerability assessment an the Mediterranean. In: Regional Activity Centre for the Priority Actions

Programme. Split, Croatia: Regional Activity Centre for the Priority Actions Programme; 2010

[34] Beller EE, Spotswood EN, Robinson AH, Anderson MG, Higgs ES, Hobbs RJ, et al. Building ecological resilience in highly modified landscapes. Bioscience. 2019;**69**:80-92. DOI: 10.1093/biosci/biy117

[35] Mock G, Salvemini D. Assessing Landscape Resilience: Best Practices and Lessons Learned from the COMDEKS Programme. New York: United Nations Development Programme; 2018

[36] Burkhard B, Kandziora M, Hou Y, Müller F. Ecosystem service potentials, flows and demands – Concepts for spatial localisation, indication and quantification. Landscape Online. 2014; **34**:1-32. DOI: 10.3097/LO.201434

[37] Berghöfer A, Mader A, Patrickson S, Calcaterra E, Smit J, Blignaut J, et al. TEEB Manual for Cities: Ecosystem Services in Urban Management. United Nations Environment Programme; 2011

[38] Kosztra B, Büttner G, Hazeu G, Arnold S. European Environment Agency. Updated CLC Illustrated Nomenclature Guidelines. Wien, Austria: European Environment Agency; 2019

[39] Sowińska-Świerkosz B, García J. What are nature-based solutions (NBS)? Setting core ideas for concept clarification. Nature-Based Solutions. 2022;**2**:100009. DOI: 10.1016/j.nbsj.2022.100009

[40] Moosavi S. Design experimentation for nature-based solutions: Towards a definition and taxonomy. Environmental Science & Policy. 2022;**138**:149-161. DOI: 10.1016/j.envsci.2022.10.004

[41] Mell I, Scott A. Definitions and ICE Manual of Blue-Green Infrastructure. Institution of Civil Engineers; 2023. pp. 3-22

[42] Liason. European Rural Innovation Ambassadors, Milano Porta Verde. Available from: https://liaison2020.eu/abmassadors/milano-porta-verde/ [Accessed: Mar 28, 2023]

[43] Zanzi A, Vaglia V, Spigarolo R, Bocchi S. Assessing Agri-food start-ups sustainability in Peri-Urban agriculture context. Land. 2021;**10**(4):384:1-20. DOI: 10.3390/land10040384

[44] Zanzi A, Andreotti F, Vaglia V, Alali S, Orlando F, Bocchi S. Assessing ecosystem services and job opportunities in Peri-urban agriculture start-up projects. In: Diehl JA, Kaur H, editors. New Forms of Urban Agriculture: An Urban Ecology Perspective. Singapore: Springer Nature Singapore; 2021. pp. 229-248

[45] Zanzi A, Andreotti F, Vaglia V, Alali S, Orlando F, Bocchi S. Forecasting agroforestry ecosystem services provision in urban regeneration projects: Experiences and perspectives from Milan. Sustainability. 2021;**13**(5):2434:2-18. DOI: 10.3390/su13052434

[46] Chiaffarelli G, Vagge I. Methodological proposals for addressing agroecological design in periurban areas: A case study in the edges of Milan (Italy). AGROFOR. 2021;**6**:45-56. DOI: 10.7251/AGRENG2101045C

[47] Dal Borgo AG, Gambazza G, Garda E. Luoghi e Comunità: Storie di Rigenerazione. Milano Udine: Mimesis; 2021

[48] Dal, Borgo AG, Capocefalo V, Chiaffarelli G. Il ruolo delle pratiche di agricoltura rigenerativa nella produzione di servizi ecosistemici e socio-territoriali nell'area sudest di Milano, Geografia e

cibo, Nuova serie, 20. Firenze, Italia: Società di Studi Geografici; 2022

[49] Longo A. OpenAgri, 18 progetti x 30 ettari: Un masterplan per un Parco della Sperimentazione Agroecologica, Workshop di lavoro svolto dal gruppo wp7 nell'ambito del progetto OpenAgri, bando Europeo Urban Innovative Action (UIA). Milano: Dipartimento DAStU - Politecnico di Milano, Dipartimento ABC - Politecnico di Milano, Dipartimento ESP - Università degli Studi di Milano; 2018

[50] Geoportale Regione Lombardia. Available from: www.geoportale.regione.lombardia.it [Accessed: Oct 10, 2022]

[51] Rivas-Martínez S, Sáenz S, Penas A. Worldwide bioclimatic classification system. Global Geobotany. 2011;**1**(1):634

[52] Prodromo della Vegetazione Italiana. Available from: www.prodromo-vegetazione-italia.org [Accessed: Oct 21, 2022]

[53] Fahrig L, Girard J, Duro D, Pasher J, Smith A, Javorek S, et al. Farmlands with smaller crop fields have higher within-field biodiversity. Agriculture, Ecosystems & Environment. 2015; **200**:219-234. DOI: 10.1016/j.agee.2014.11.018

[54] Wu J, Hobbs RJ. Key Topics in Landscape Ecology. Cambridge, UK: Cambridge Studies in Landscape Ecology; 2007

[55] LOS_DAMA! landscape open space development Alpine metropolitan areas, EU-Interreg V B project. LOS_DAMA! Compendium Green infrastructure for better living. Available from: https://www.alpine-space.eu/project/los_dama/

[56] Corona Verde Project. Available from: https://www.gioia-gibelli.it/portfolio/los_dama/

[57] Bagliani F, De Biaggi E. Corona Verde: un sistema di parchi per l'area metropolitana torinese. Ri-Vista. Research for Landscape Architecture. 2015;**1**:86-99. DOI: 10.13128/RV-17548

[58] Prat A, Mangili S, Saraco R, et al. Corona Verde 2025. Un Progetto del Terzo Piano Strategico per Mettere a Valore L'infrastruttura Verde Dell'area Metropolitana Torinese. I quaderni di Torino Strategica 1. Associazione Torino Internazionale; 2016. Available from: http://www.torinostrategica.it/wp-content/uploads/2016/06/Quaderno_Verde_web.pdf

[59] Cesarini C et al. Piano Territoriale Metropolitano. Documento di Sintesi. Città Metropolitana Milanese. 2022. Available from: https://www.cittametropolitana.mi.it/export/sites/default/PTM/iter/PTM_pubblicato_burl/media/PTM_DocumentoDiSintesi_20220321.pdf

[60] PTM Milano. Adattarsi ai cambiamenti climatici. Available from: https://www.gioia-gibelli.it/portfolio/ptm-milano/

[61] Gibelli MG. 2011. Scalo Intermodale. In: Monti AL, Villa P, editors. Architettura del Paesaggio in Italia. Modena: AIAPP. Logos; 2011

[62] Gibelli MG. Acque e città: pianificare con l'acqua. Biologia Ambientale. 2017; **31**:253-259

[63] Gibelli MG. Nature Based Solutions ieri, oggi e domani. Reticula. 2021;**28**:5-12

[64] F.A.R.E. ARCO VERDE. Favorire l'Attuazione della Rete Ecologica: primi interventi di concretizzazione delle proposte progettuali individuate dallo Studio di fattibilità Arco Verde. Provincia di Bergamo; 2015. Available from: https://82.115.177.2/arcoverde/infodiscs/index/1.html

**Chapter 5**

# Mesoamerican Pyramidal Shaped Landscapes as Maps in Urban Ceremonial Sites: "Close Mapping" Analysis of Spatial Homologs

*Dennis Doxtater*

## Abstract

Probability tests show the likelihood of designed and land-surveyed "landscape pyramids" laid out horizontally over large-scale geographic areas. Linked are two coincidental equinox and solstice alignment patterns among volcanos at Mexico (valley) and Lake Atitlan. Pattern apexes position early Olmec sites of La Venta and San Lorenzo. This frame provides "benchmarks" to locate and orient site features for both Olmec and early Mayan foci of Piedras Negras, Palenque, El Mirador, and Tikal as exemplary. The design of these complex ceremonial sites is yet to be understood, despite attempts by notable applications such as Space Syntax. The present work uses custom software to accurately map formal geometry at large landscape scales, seeking azimuthal homologs (identities) of individual alignments and pyramid-shaped patterns with urban scale designs. Homolog patterns will serve future research to statistically prove intentional design. Hypothetically, homologic mapping served to integrate diverse social groups and their ceremonial foci over large geographic areas, unifying ritual-based religion and diminishing territorial strife.

**Keywords:** homologs, prehistoric land surveying, Mesoamerica, ritual landscapes, urban design

## 1. Introduction

A possibly "primordial" religious concept in Mesoamerica mapped a coincidental, topographic area of prominent volcanos south of Cerro Gordo (and eventually Teotihuacan just below), creating a horizontal sacred pyramid shape, **Figure 1**. Ethnographies of "cosmic" ritual in Mesoamerica can be found in Refs. [1–6].

This idea assumes that early Mesoamerican people had the technical surveying ability to understand the unique accuracy of geometric elements of this pattern. Prominent is the position of the highest volcano of Orizaba in winter solstice alignment with the apex at Cerro Gordo (*via* interim station point), an accurate equinox

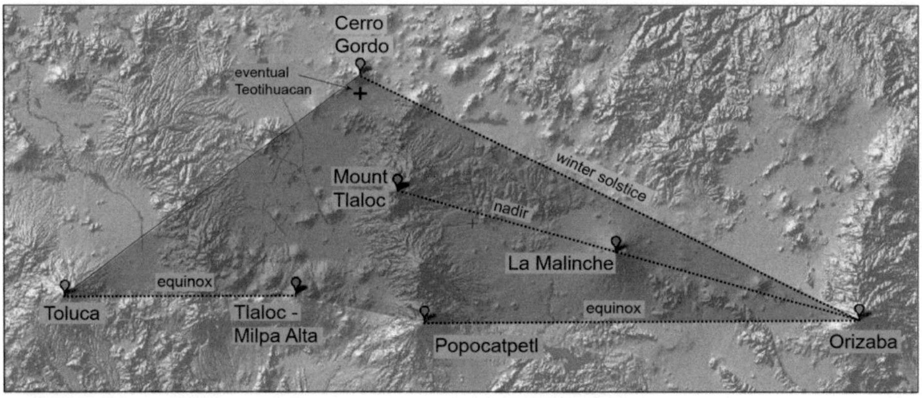

**Figure 1.**
*Coincidental pyramid pyramid-shaped association of solstice, equinox, and zenith nadir azimuths from Cerro Gordo azimuth. Photo of pyramid shaped shadow of highest mountain in Mesoamerica, Orizaba (open-source blog in peterzaitsev.com, Jan. 05, 2023). Šprajc's [7] archaeoastronomy analysis of Teotihuacan's site orientation does not involve any of these features except for the immediately visible Cerro Gordo. Alternatively, see Ref. [8].*

alignment west to Popocatpetl, the second highest volcano, and the Nadir sunrise from the archaeological feature on the top of Mount Taloc [9] with the peaks of La Malinche and Orizaba. Equinox alignment from Toluca to Tlaloc – Milpa Alta forms the complementary western baseline of the pyramid.

Cerro Gordo's horizontal "landscape pyramid" may have been emulated when the Olmec positioned La Venta. Symbolically, this pivotal site mediated a great diagonal azimuth between Cerro Gordo, the volcanos around Lake Atitlan, and the far south volcano terminus of San Salvador. As drawn in **Figure 2**, the La Venta landscape pattern replicates two sets of equinox-aligned mountains, the line between the two inner points creating the intersection for an accurate meridian to La Venta. Additionally, the intersecting diagonal from Orizaba to the volcano San Salvador precisely runs through Atitlan X, the right angle connecting the highest southern volcano, Tajamulco, and the storied Atitlan/Toliman meridian pair. This earliest possible pattern emulation was tested against random probability. Prehistoric surveying at large scales requires many station points, each within sighting distance of the two adjacent stations. Through iterations of aligning threesomes sequentially back and forth along the entire length of the line, accuracy down to about 0.02° can be achieved (see chapter in Refs. [11–13]).

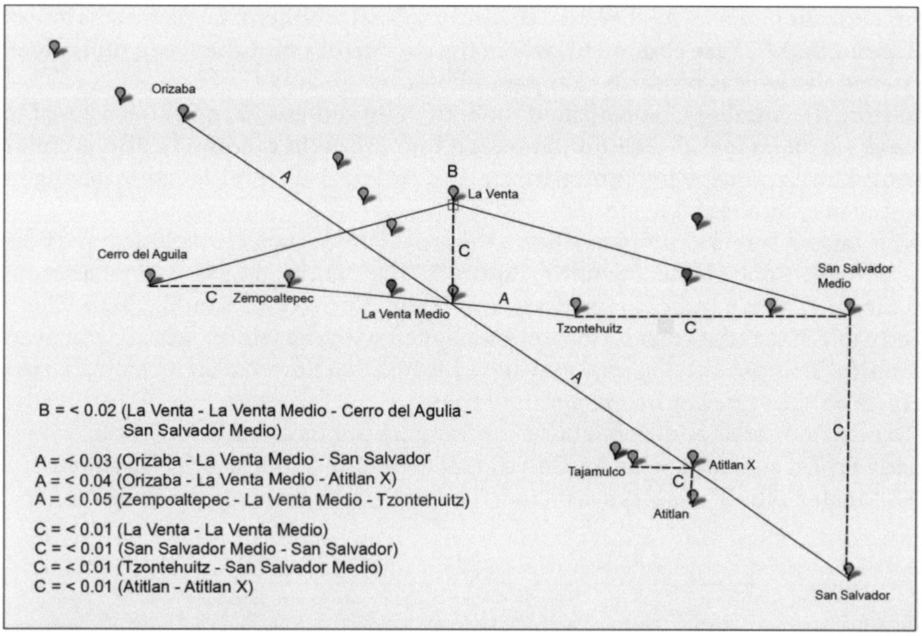


Patterns Investigated (1):

A+A(2)+A(1)+C(1)+C(1)+C(1)+C(2)+B(3)

1: set 0 Alignment: orizaba , la venta medio , atitlan x , 235.657843 Alignment: orizaba , la venta medio , san salvador , 235.697709 Alignment: zempoaltepec , la venta medio , tzontehuitz , 264.490349 Cardinal: la venta , la venta medio , 179.992167 Cardinal: atitlan , atitlan X , 000.0000 Cardinal: san salvador medio , tzontehuitz , 89.999921 Cardinal: san salvador medio , san salvador , 179.999880 Bisect: la venta , la venta medio , san salvador medio , cerro del agulia , 75.008412 , 74.984822 id=1

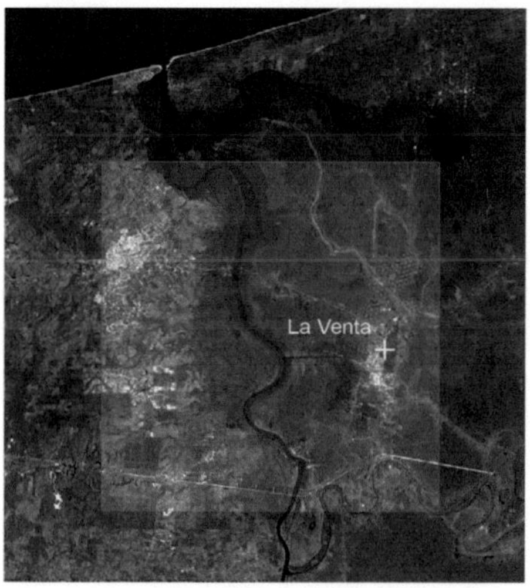


**Figure 2.**
*Screenshot from probability analysis of existing La Venta location compared to random points positioned in test box (photo left); patterns investigated box shows how search string finds existing total pattern; site background in Ref. [10].*

**Figure 2** shows test data from substituting random points within a defined box around La Venta. Similar testing for design probability can be evaluated in Refs. [14–20]. Presently, how often does a random point reproduce the existing pattern defined by the search string shown in the "Patterns Investigated" box? Using existing accuracies, alignments (A), bisects (B), and cardinals (either true meridians or east-west azimuths) (C) are combined, where the overlap of points between individual geometric elements is proscribed in parenthesis.

Setting the number of substituted random point sets at 100, no pattern equal to the existing one is found. At 1000, however, four different random locations create an equivalent landscape pyramid pattern. The odds of La Venta's location having been designed may be about 4 in 1000 or 0.004.

The pattern is more complex when additional Olmec sites are included, particularly the huge Aguda Fenix complex, **Figure 3** [21]. In this possible design, a second long diagonal runs parallel to the first, connecting Cerro Gordo and La Malinche directly to Tacana, the volcano viewpoint aligned with the winter solstice rise over Tajamulco. This second diagonal may have located San Lorenzo on its meridian to the intersect on the same landscape pyramid base.

To test, one draws additional boxes for random points around San Lorenzo, Aguada Fenix, and Laguna de los Cerros [22]. The core pattern of "N" is added for 90° angles. No "B" bisects are included because of their (counterintuitive) far

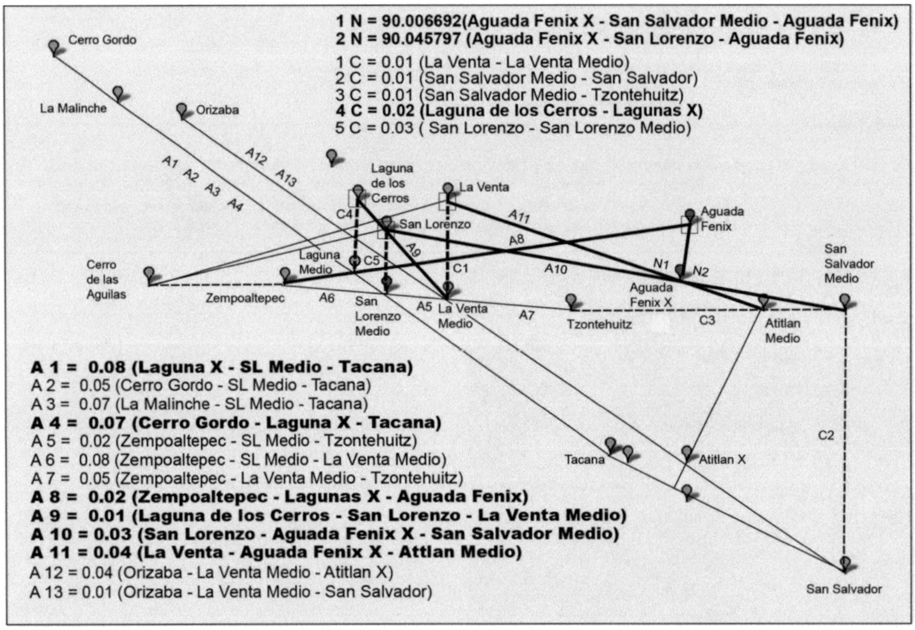


Patterns Investigated (1):

A+A(2)+A(2)+A(3)+A(1)+A(2)+A(3)+A(2)+A(1)+A(2)+A(1)+A(1)+A(1)+N(3)+N(3)+C(2)+C(2)+C(2)+C(2) (1)

1: set0 Alignment: lagunas X , sl medio , tacana , 234.003013 Alignment: cerro gordo , sl medio , tacana , 234.126385 Alignment: la malinche , sl medio , tacana , 234.149527 Alignment: cerro gordo , lagunas X , tacana , 234.230018 Alignment: zempoaltepec , sl medio , tzontehuitz , 264.708646 Alignment: zempoaltepec , sl medio , la venta medio , 264.769726 Alignment: zempoaltepec , la venta medio , tzontehuitz , 264.490349 Alignment: zempoaltepec , lagunas X , aguada fenix , 278.636137 Alignment: orizaba , la venta medio , atitlan X , 235.657843 Alignment: orizaba , la venta medo , san salvador , 235.697709 Alignment: laguna de los cerros , san lorzenzo , la venta medio , 222.646751 Alignment: san lorzenzo , aguada fenix X , san salvador medio , 259.964943 Alignment: la venta , aguada fenix X , atitlan medio , 250.244953 Ninety: aguada fenix X , san lorzenzo , aguada fenix , 90.045797 Ninety: aguada fenix X , aguada fenix , san salvador medio , 90.006921 Cardinal: la venta , la venta medio , 179.992167 Cardinal: san salvador medio , tzontehuitz , 89.999921 Cardinal: san salvador medio , san salvador , 179.999880 Cardinal: laguna de los cerros , lagunas X , 180.010366 id=1

**Figure 3.**
*Screenshot of a more complex probability test using additional test boxes at San Lorenzo, Aguada Fenix, and Laguna de los Cerros; the result shown is for the existing pattern.*

greater occurrence in complex patterns. Evaluating 1000 sets, no matches are found. Likewise, at 10,000. Generating 100,000 sets, 4 fit the search string (though with minor variations at this complexity). The second run of 100,000 finds 2 matches, and in the third run of 100,000 only 1 occurs. Thus, the probability of the overall pattern being randomly produced is 7 in 300,000 or 0.000023. A separate exercise positioned two alternative meridian locations of only La Venta, north and south of the actual site. Calculating the pyramidal geometry of these two apex test points showed a clear precision for the existing location.

Probability testing in cultural landscape research needs critical academic discourse as does the inescapable likelihood that these societies could technically map large-scale landscape coincidence and develop integrative design. This research should logically follow radically new scales of investigation created by lidar mapping from the air. In a recent *Science* article describing the massive scale of urbanization in Amazonian Ecuador, the authors state:

> *"Indeed, the overall organization and standardization of architectural features, the road patterns, or the defense systems, suggests, for instance, the existence of advanced engineering (mature architectural tradition relying on sighting and surveying instruments and/or expertise)" [23].*

## 2. Individual azimuths as homologs: piedras Negras development in context with La Venta, Lagunas de los Cerros, Tres Zapotes, and Aguada Fenix

Analyzes of historic or prehistoric architecture/urban design form conventionally flows from critical art theory tracking rhetoric of artists. The most successful methodology of understanding more abstract meanings of social space developed under the rubric "Space Syntax" [24]. This proprietary computer methodology maps patterns of human use in defined architectural or urban form, discovering existing or potential territorial (accessible or inhibiting) behaviors. As such, a number of archaeologists have so-mapped traditional or prehistoric sites, identifying egalitarian or hierarchical implications of the form (several in SW Pueblo culture, e.g., the dissertation of Ferguson [25]). The question remains, however, how to approach more distinctly symbolic settings and their ritual practice. Geometries of alignment, threshold, opposition, symbolic domain, and especially homolog, not to mention technical precision of layout, are from quite a different toolkit. Archaeologists Morton et al. [26] recognize as much, concluding their Space Syntax analysis of portions of one of the most religiously powerful Mesoamerican urban sites of Teotihuacan:

> *It may be suggested that comparing commoner households and civic-level planning, presumably by an elite class, is like "comparing apples and oranges". We wish to make emphatically clear that we do not suppose that these clearly different socio-spatial categories should be referential one to the other. Rather, we are suggesting that, in this case, whether intentional or not, these different levels are symbolically referential. To this extent, while we are able to offer no concrete conclusions, we will end this article with a brief discussion of some of the potential consequences of this relationship for community identity and civic authority in epicentral Teotihuacan and explore this model within a broader context of early urbanism [26].*

The present paper offers a new means of analyzing the formalized, ritually used geometry of "civic-level" planning as distinct from behaviors of "commoner households." Going beyond probability testing of large-scale patterns, analysis of closely mapped azimuths at landscape and urban scales discovers homologies (replication of patterns) of individual alignments and horizontal landscape pyramid patterns.

Study of large-scale design formalities did not begin with any layperson "discovery" using the minimal geometric mapping afforded by Google Earth. Rather, the author, an architect and socio-cultural anthropologist, over 40 years ago began collecting any ethnographic reprint with plan illustrations of ritually used spaces; almost all were at small spatial scales showing formal geometries in the layout of traditional dwellings and settlements [27–32]. Eventually, the possibility of prehistoric technical ability to land survey—through fieldwork exercises and articles on the subject, especially in Egypt and Rome—raised dramatic new questions about possible scales of ritual frameworks. Eventually, a software company with a math background built a custom app that could quickly define formal properties between points on an elliptical earth's surface (NOAA accuracies) and test existing patterns against random ones [13].

Hypothetically, after laying out the two "landscape pyramids" of San Lorenzo and La Venta, the remarkably coincidental location of the eventual site of Piedras Negras created an alternation to La Venta geometry. **Figure 3** focuses on the intersection point of the two lines running from the two landscape pyramid apexes down to their two respective base points on the equinox running east from Tzontehuitz. The San Lorenzo line of this intersection forms a right angle from the intersection to the center of the massive Aguada Fenix. While site points in Aguada Fenix and Piedras Negras can be relatively accurately determined, the lack of any architecturally well-defined map of San Lorenzo makes this point more speculative (site background in Ref. [33]). Estimating from the rough topographic remains at San Lorenzo, the right angle measures about 90.21955°. More accurate in **Figure 4** is the opposed right angle from Aguada Fenix to Piedras Negras and a better-defined point at Tres Zapotes, that is, 89.9066°. The two azimuths west from the north and south points of Piedras Negras – Aguada Fenix are accurately parallel.

**Figure 4** also shows the formally massive ceremonial space at Aguada Fenix. As a clearly designed orientation, this 1400-meter airstrip-like plaza may be less inspired by some Pan-Mesoamerican astronomy precedent, as briefly mentioned in this publication—than a land surveyed relation to the two right angles to Piedras Negras and Tres Zapotes. It orients about 170° to the south while the azimuth to the Piedras Negras point is 168.78605 to the ball court in the north and 169.91007° to the center of the south court. Furthermore, the coincidental location of the Piedras Negras may have altered La Venta's original landscape pyramid pattern,

Geometrically, Piedras Negras has a Western partner, the well-known monumental site of Monte Alban, as evidenced by the closely equal distances of their centers to La Venta, also shown in **Figure 4**. This could have necessitated a revision of La Venta's central meridian to a new point canted eastward on the equinox base; note the equal angles to this point from the two opposed vertex sites. Hence, this is the orientation of La Venta's overall site design. This new base point also works to establish two additional Olmec sites at Lagunas de los Cerros and Tres Zapotes.

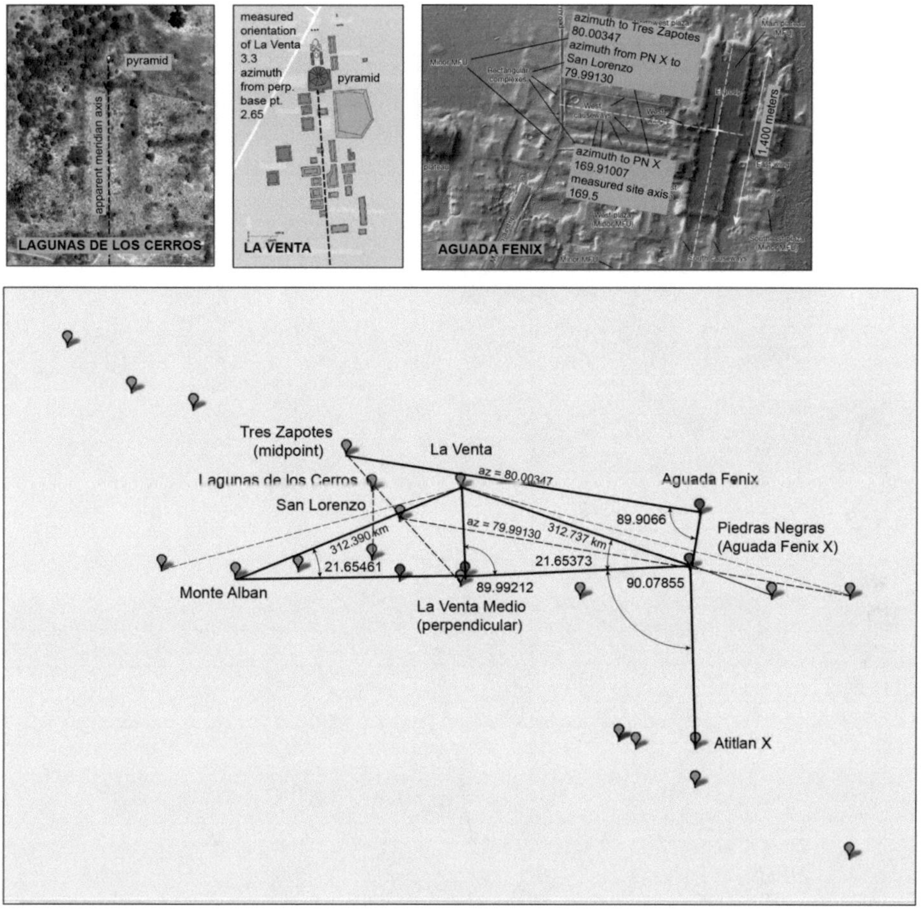


**Figure 4.**
*Possible revision of La Venta's meridian down to the equinox pyramid base: Isometric angles and equal ray distances from Monte Alban and Piedras Negras to La Venta, and its new base location that may have created site orientation. Aguada Fenix's site axis and perpendicular as expressive of right-angle azimuths to Tres Zapotes and Piedras Negras. Lagunas de los Cerros location and orientation on accurate meridian from point on alignment from Zempoaltepec to Aguada Fenix (**Figure 3**). Alignment of Tres Zapotes, San Lorenzo, and original La Venta base point.*

## 3. La Venta's landscape pyramid as map in Piedras Negras

The design rationale of the geometrically complex site plan of Piedras Negras [34–37] has never been deciphered, despite the inspired but limited efforts by archaeoastronomer Hartung [38]. Initiating this massive ceremonial center, a founding axis might have followed the azimuth from Aguada Fenix down to the best large-scale point at the center of the eventual South Group Court, **Figure 5**. This azimuth from the court center runs accurately through one of the largest temples, "O-12."

Consider again how Aguada Fenix's right angle prolongs down to invoke the coincidental San Lorenzo and La Venta east rays intersection point that is Piedras Negras. Using the center of the eventual South Court, the Aguada Fenix azimuth to

**Figure 5.**
*La Venta landscape pyramid with apex at great stair feature of Piedras Negras; major site meridian runs through apex point; parallel azimuths from apex and K-5 north temple determine acropolis high point feature, azimuth to ball court center is San Lorenzo Medio (meridian base point for its landscape pyramid) to Piedras Negras.*

this point might have first created the interim location of the O-12 temple, **Figure 5**. For its part, O-12's temple orientation does not mimic the Aguada Fenix azimuth, but the original La Venta landscape pyramid east equinox vertex running from Atitlan Medio to Atitlan X. This site scale homolog from O-12 runs down to the R-16 pyramid north of, but not part of the South Group Court. The R-16 temple contributes to a major site meridian connecting opposed domains of Piedras Negras; temples K-5, R-5, and R-4 may also align accurately with this true north-south axis. Most conceptually intriguing is the meridian position of a monumental, logically ritually powerful stair mediating the two undoubtedly symbolic domains.

While enlarged on-site dimensions will be required to refine presently drawn design concepts, the meridian passing through the formally dominant center stair of Piedras Negras may have expressed the apex of the original La Venta landscape pyramid meridian, suggesting a large-scale geographical area appropriated as small-scale homolog in the layout of a late Olmec or early Mayan ceremonial precinct. Tentatively, the azimuth from the center of the monumental stair to the ritual top of O-12 mimics the original east ray of La Venta, that is, from its apex to the base vertex at Atitlan Medio. Then, when calculating La Venta's west ray to Cerro Aguila, this azimuth runs to the smaller feature of N-1. The angle from N-1 to the O-12 point is cardinal, as would have symbolically evoked the equinox meaning of the two original base vertexes of the La Venta landscape pyramid.

Also interesting in **Figure 5** is the highly symbolic location of K-5 sitting on the northern apex of Piedras Negras' major organizing feature. Did it emphasize the aboriginal power of La Venta and its meridian as a mediator of ritual passage between western and eastern volcano domains *via* the great diagonals? Later, perhaps, the acropolis as a high point of the Western domain becomes more powerful as it integrates the La Venta landscape pyramid homolog. The extension of the ray from O-12 through the "great" stair runs to the high point of the acropolis, perhaps emulating La Venta's apex power. The azimuth from the acropolis high point to the K-5 meridian temple mimics the western ray of the original La Venta landscape pyramid, Cerro Aguila – La Venta. Does this express social-symbolic linkage to the west above Cerro Aguila, perhaps even Cerro Gordo and Teotihuacan?

Two additional azimuths suggest the integration of the ball court in front of K-5 with the central Piedras Negras stair homolog. First, the line from the acropolis high point to the ball court center runs through the largest temple of the acropolis. The azimuth of this line matches that from San Lorenzo Medio (where its meridian strikes the great diagonal from Cerro Gordo, **Figure 3**) to an early Piedras Negras intersection point. The second line, necessary to locate and possibly orient the northern ball court, runs from the western base of the homolog, N-1, up to the apex of the meridian, K-5. This azimuth may be perpendicular to the La Venta great diagonal from Orizaba through Atitlan X to San Salvador. Thus, both great diagonals might be symbolized in the location of acropolis features and ball court, both integrating with the original concept of La Venta as a landscape pyramid.

An alternative interpretation of K-5 on the meridian defines this symbolically powerful location and temple as politically created by people from Aguada Fenix. After all, the location of Aguada Fenix may have originally been determined by the coincidental intersection of the two east rays from San Lorenzo and La Venta—again reflecting a powerful integration of the two great diagonals and their meridians that defined the early Olmec landscape. If Aguada Fenix founded Piedras Negras, then it makes sense for its central stair to symbolically point north to K-5 as Aguada Fenix.

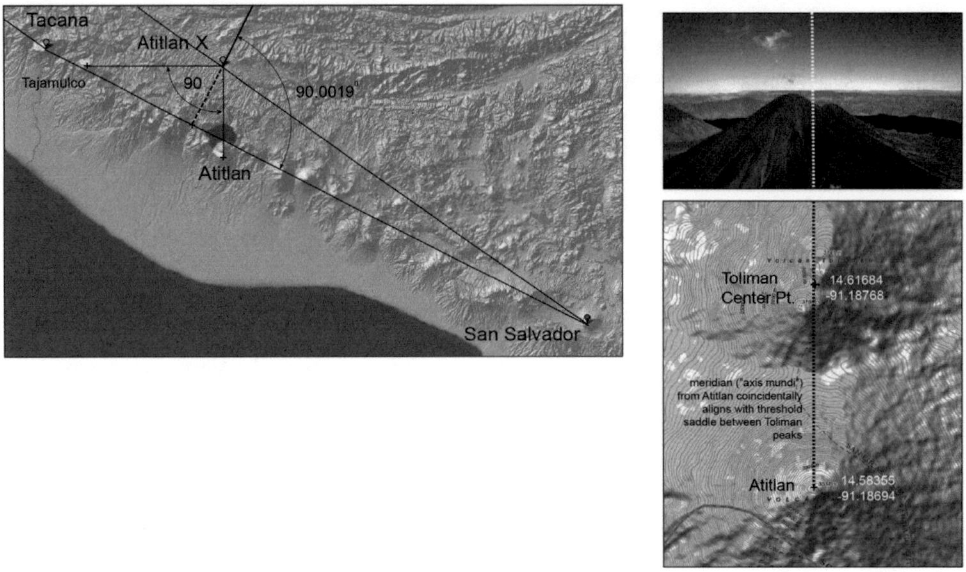


**Figure 6.**
*Unification of two great diagonals at Atitlan X, which terminate at Tacana and San Salvador: Coincidental meridian through Atitlan and Toliman Center Pt. align with Atitlan X and form right angle to highest mountain, Tajamulco; non-coincidental line between Tacana and San Salvador serves as base for designed perpendicular through Atitlan X that runs north to create east vertex point on base of formative La Venta landscape pyramid.*

Considering the opposed temple south of the stair on the meridian, the azimuth from O-12 again may be from Atitlan Medio to Atitlan X, **Figure 5**. The symbolic logic of Atitlan Medio and Atitlan X can be better understood in **Figure 6**.

## 4. Palenque's homolog landscape pyramid

The well-publicized dramatic, religious architecture and history during the classic (AD) period at the Mayan site of Palenque [39] makes no attempt to map the origins of the ceremonial city in its earliest BC period. Now, considering its location in the formalized Olmec landscape, an emulation of the original La Venta pattern makes sense, **Figure 7**. Notably, however, Palenque assumes the most ritually powerful vertex position of the landscape pyramid, perhaps because of its dating several centuries later than Olmec sites and even Piedras Negras, suggesting a diminished power of earlier patterns.

Nevertheless, a Palenque landscape pyramid appropriates the original La Venta equinox base points of its meridian, La Venta Medio (to Tzontehuitz), and its eastern Atitlan Medio connecting down to Atitlan X. In **Figure 7**, the angles from these two base points to the Palenque vertex are close to identical: 21.05908° and 21.03561°. Emulating Olmec landscape pyramids, Palenque, too, has an accurate north-south meridian down to the early Mayan point of Izapa's southern plaza, which forms angles of 72.04227° and 72.06378° from Palenque vertex to base points.

While the azimuth of the east ray of this landscape pyramid does *not* run to La Venta, like Piedras Negras, designer priests may have paid homage by so orienting the dominant central temple ("palace") and ritually significant ball court (perpendicular). Most symbolically interesting to archaeologists, however, has been the Temple of

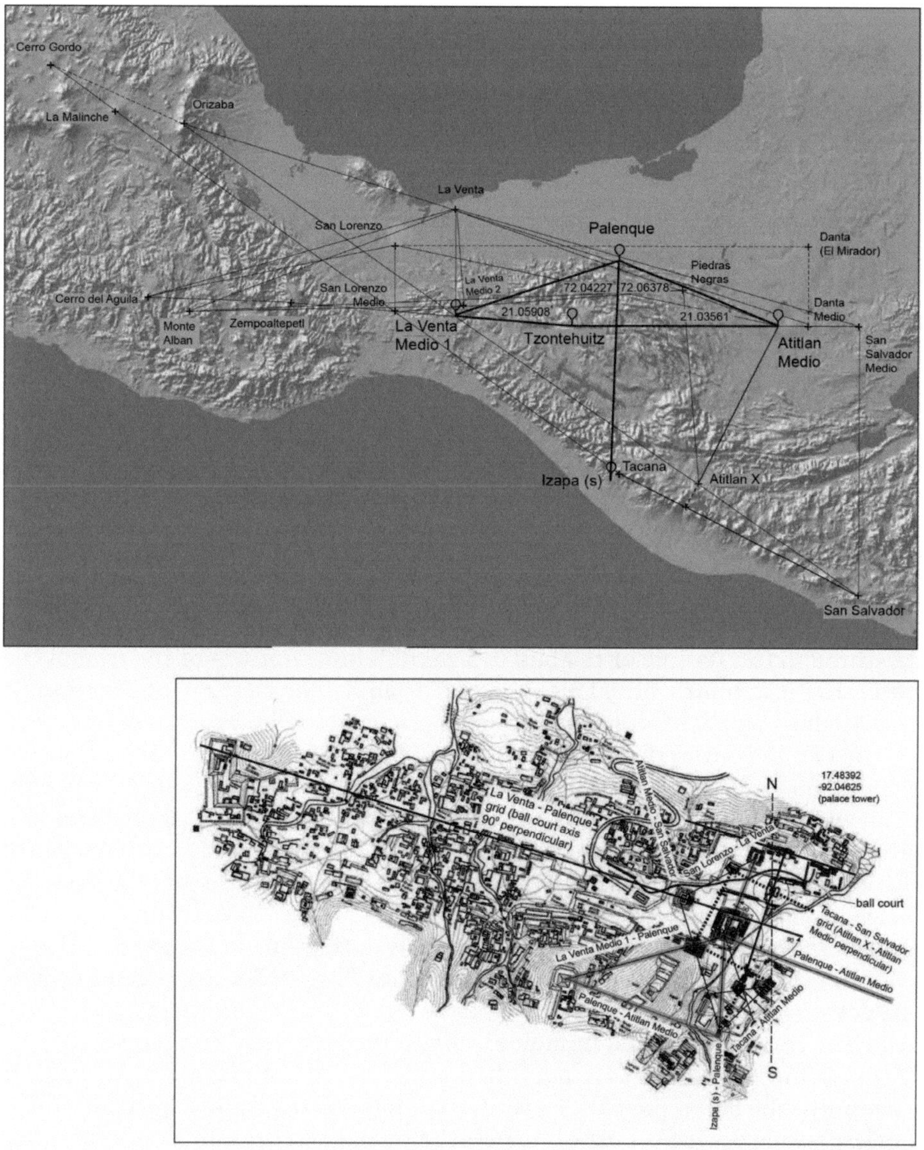


**Figure 7.**
*Alignment of Palenque's central axis to La Venta; equilateral vertex angles formed with two original equinox landscape pyramid base points, La Venta Medio and Atitlan Medio; vertical axis to Izapa creates equal angles and aligns with central palace and inverted triangle apex.*

the Cross group to the south of the palace, **Figure 8**. Its Temple of the Sun may exist as the focus of three temples and their large bases. The Temple of Inscriptions, Temple of the Cross, and Temple XX form a triangle whose azimuths can be evaluated as part of a site homolog to a larger formalized landscape, **Figure 9**.

First, the major vertical axis of the triangle running through the threshold of the Temple of the Sun could have been the azimuth from Atitlan X to Atitlan Medio, again an expression of the Piedras Negras intersect (La Venta ray) and the fusion of the two great diagonals. Also involving Atitlan X, the line from the palace

**Figure 8.**
*Temple of the Cross (Wikipedia open access).*

center through the Temple of the Sun to a small temple outside of the triangle corresponds to the azimuth from Palenque to Atitlan X. The west ray of the triangle is Atitlan Medio – San Salvador. The triangle's opposite ray is Tacana – Atitlan Medio. Thus, the two rays from the triangle's southern Temple XX vertex unify lines from Atitlan Medio (La Venta base) to the two ends of the great diagonals (Tacana and San Salvador). Logically, the azimuth between these two great diagonal end points superimposes at Palenque between the Temple of Inscriptions and the Temple of the Cross. The accurate azimuth at the top of the triangle runs from the back top of Inscription to front top of Cross. Archaeo-astronomers have studied the solstice relation between the two temple tops (shown as dotted line in **Figure 9**). Also apparent in the triangle layout is a meridian from Temple XX to the center point of the palace.

All these lines correspond homologically to the large-scale map inset in **Figure 9**. While map-like with individual lines true to their large-scale location and azimuth, the Palenque pattern is rotated both vertically and horizontally. Did priest surveyors have a list of principal azimuths in the Olmec toolkit, and could replicate these orientations accurately? Triangle designers at Palenque might have first recognized the ultimate power of the Tacana-San Salvador line that not only unified the two great diagonal termini (to Cerro Gordo and Orizaba), but its perpendicular that passed through Atitlan X on its way to mark the east equinox landscape pyramid vertex, Atitlan Medio, of the La Venta pattern. Symbolically, the Temple of Inscriptions is "Tacana" and the Temple of the Cross "San Salvador." Locating both volcanos close to the Palenque center, rather than the abstract benchmark of Atitlan Medio (Temple XX), makes symbolic sense, perhaps explaining the vertical inversion. Given the intention of homologizing the of Tacana, San Salvador, and their integration by Atitlan Medio (the founding east triangle base of La Venta), it may have been logical to position Temple XX using the known azimuths of the two rays in question but inverting them horizontally. This clear expression of Atitlan Medio can then be aligned with the all-important

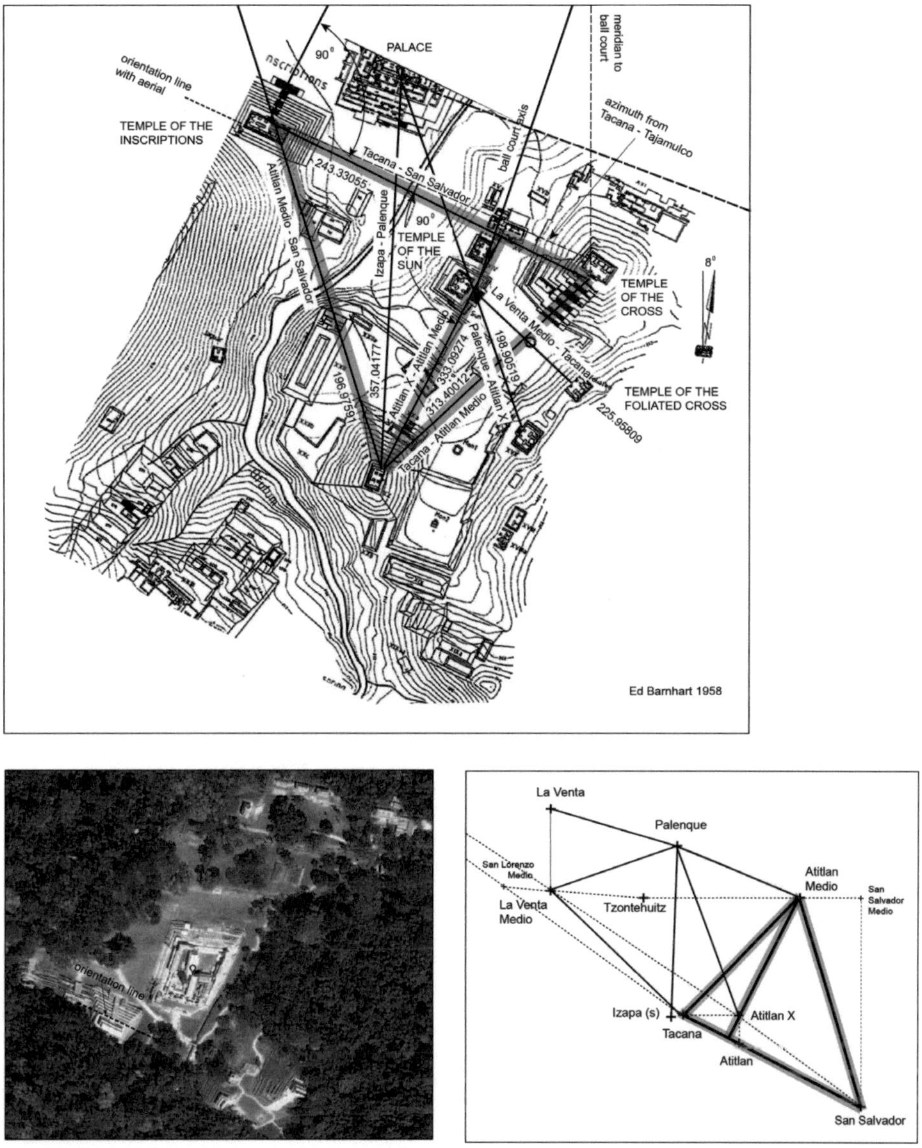


**Figure 9.**
*Photo shows means of orienting published site plan. Inverted and rotated large-scale map (below) possibly reproduced as homolog in layout of southern portion of Palenque; interior axis symbolically expresses spiritual power of Atitlan X's unification of founding great diagonals in relation to La Venta and meaning of the Temple of the Sun to the Cross group.*

axis in the triangle from Atitlan X, and the Cross Group focus of the Temple of the Sun. Though perhaps not the present case, Mayan priests understood "flipped" map orientations of patterns transformed homologically from the sky to ceremonial sites (discussions in Ref. [40]). Despite the symbolic inconsistencies, Palenque designers seem likely to have clearly conceptualized the large-scale triangular landscape pyramid map as they laid out the southern portion of the site.

## 5. Olmec patterns in the location and map homologs of La Danta and El Mirador

One of the world's most massive architectural pyramids, La Danta, looms on the east edge of one of the largest early Maya urbanizations, El Mirador [41, 42]. Using Danta as large-scale benchmark, one sees a strategic connection to the early Olmec landscape pyramid pair *via* two cardinal right-angle azimuths, **Figure 10**. First is its close latitude, 17.75155° to that of San Lorenzo, 17.72022° (best estimated point). Second, the azimuth from Orizaba through La Venta down to the equinox base replicates the Olmec pattern of generating a right-angle meridian up to a prospective ceremonial site. The point on the equinox baseline is here labeled "Danta Medio." If extended 103.385 km precisely north, the meridian comes within 1036 meters east of the Danta pyramid top (azimuth of 0.57133°).

La Danta lies curiously well apart from the ceremonial core, its small front central feature aligning the orientation of the massive pyramid to the central feature of the Gran Acropolis Central, **Figure 11**. Not surprisingly, this axis captures the symbolic power of the two lines that may have located La Danta, that is, the opposed end points of San Lorenzo-Danta and the meridian Danta – Danta Medio. The primary site azimuth is thusly San Lorenzo – Danta Medio. The Gran Acropolis focus creates

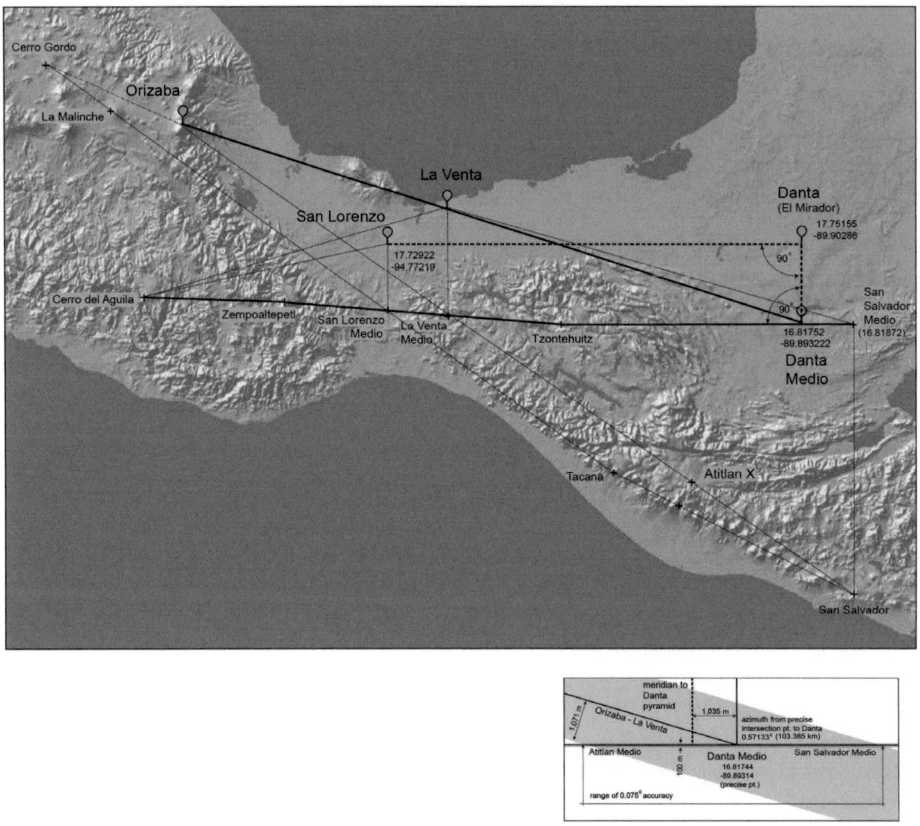


**Figure 10.**
*Accuracy of meridian from equinox base point created by alignment with Orizaba and La Venta to La Danta point, cardinal east from San Lorenzo, follows Olmec meridians at La Venta and San Lorenzo, and later examples in Ancestral Pueblo religion [8, 43, 44].*

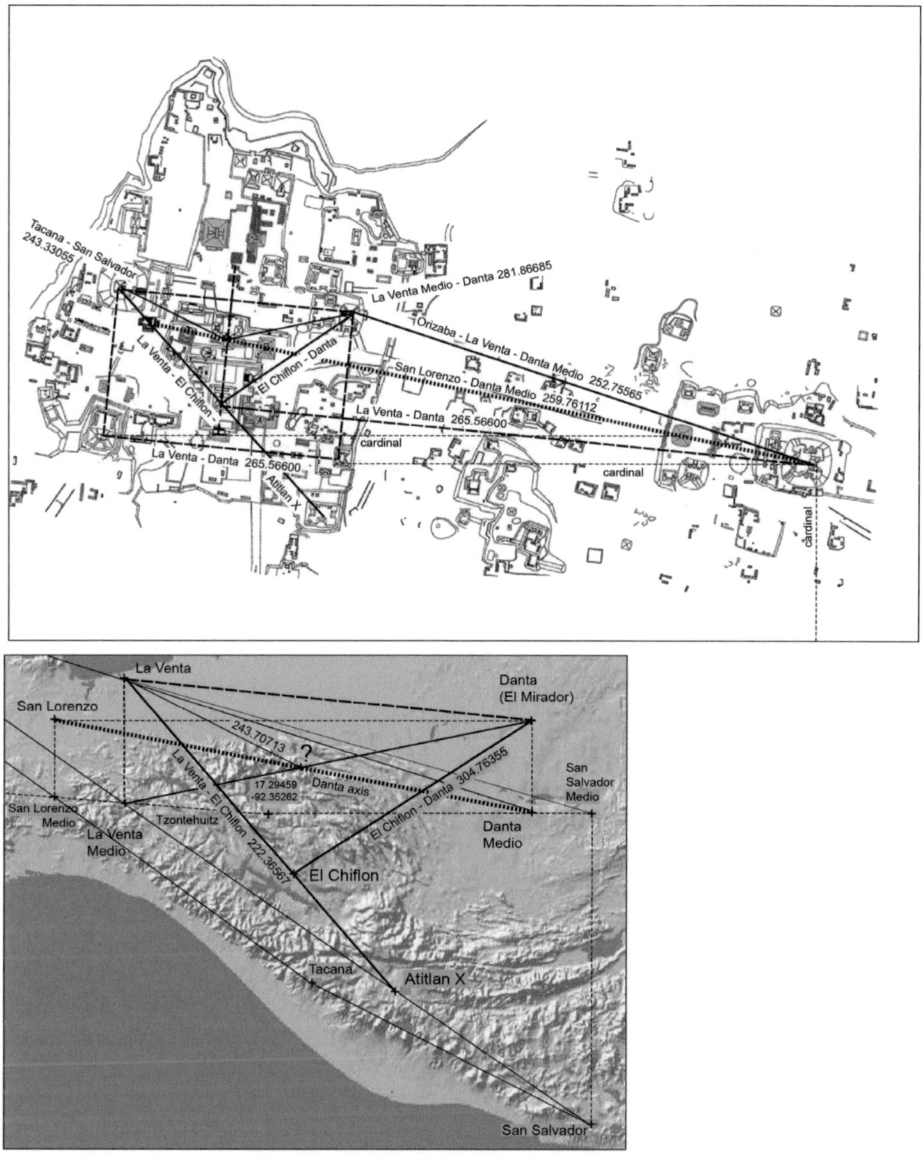


**Figure 11.**
*Homolog of azimuth from Danta Medio to San Lorenzo as principal axis at El Mirador; inverted pyramid form as correctly oriented homolog map of large landscape pattern; side axes from La Danta replicate founding azimuths of Orizaba-La Venta-Danta Medio and La Venta-Danta which also orients overall site rectangle; important site axis from El Tigre through triangle vertex expresses symbolic power of El Chiflon aligned with La Venta and Atitlan X.*

a triangle between three flanking temples, not unlike the Palenque. At El Mirador, two additional lines from Danta's summit form a bisect to the two temples of the east ray of the triangle. The azimuth of the northern ray from La Danta is that from Orizaba-La Venta-Danta Medio, while the southern azimuth is La Venta – Danta. Both run to La Venta, leaving San Lorenzo expressed by the central axis from La Danta. What large-scale map might El Mirador emulate? Looking at the two largest temples that form the upper line of the acropolis triangle, El Tigre at the western vertex, this azimuth may have expressed the most powerful meaning of El Mirador—its reverence to La Venta. All four orientations of the rectangle that surrounds the triangle, as well as the southern ray from its inverted apex to La Danta are created from this azimuth. The triangle, therefore, logically maps to the line between La Venta and La Danta, and one asks where is the triangle apex in this geography?

Following Palenque's layout, if one places the azimuth from El Tigre (La Venta) to Atitlan X on the calculated large-scale map, it fits. But the location of Atitlan X lies considerably to the southwest of the inverted triangle vertex. A unique waterfall, El Chiflón, sits on the La Venta – Atitlan X line, **Figure 12**. Before laying out central El Mirador and its triangle (after positioning La Danta), priests/surveyors would design the large-scale map to be reproduced as site layout. Neither the direct azimuth from La Venta to La Danta nor that from La Venta to Atitlan X had been previously surveyed. Teams in the field positioning themselves on intervisible high points between La Venta and Atitlan X would have discovered the coincidental alignment to the uppermost pool of the cascade. Evidence of this knowledge exists in the azimuth of El Mirador's east triangle ray that matches the large-scale El Chiflón – La Danta line.

What, then, does the central axis of these two triangles from El Chiflón express—in both map and ceremonial site? Does a site yet to be discovered exist as the homolog of the central temple of the Gran Acropolis of El Mirador? At the Gran Acropolis, azimuths from the two northern triangle vertexes to the center mimic the two azimuths of Tacama – San Salvador (west) and La Venta Medio – La Danta (east). Did surveyors also recognize the religious need to lay out La Danta's azimuth to the point on the equinox base that created the meridian to La Venta? The intersection of this line with a San Lorenzo – Danta Medio point creates a symbolic symmetry as an unknown intersection point at the center of the large-scale map.

Clearly, however, El Mirador's triangle creates a mediating threshold between La Venta and La Danta, as both cascade and Olmec linkage between two great diagonals. Looking more closely at the precise location of Atitlan X and the El Mirador-related azimuth from La Venta through El Chiflón, **Figure 13**, a possible alignment exists with the Post-Classic Mayan site of Utatlán. It was here, walking distance from the Atitlan X intersection point—the right angle between Tajamulco and Atitlán—that the precious record of Mayan creation mythology called the *Popol Vuh* was recorded [in the colonial town of Santa Cruz del Quiché a kilometer east of the Utatlán site] [45, 46].

**Figure 12.**
*El Chiflón Cascades tourist attraction in Chiapas (left), Calakmul's pyramid II (right), open access images.*

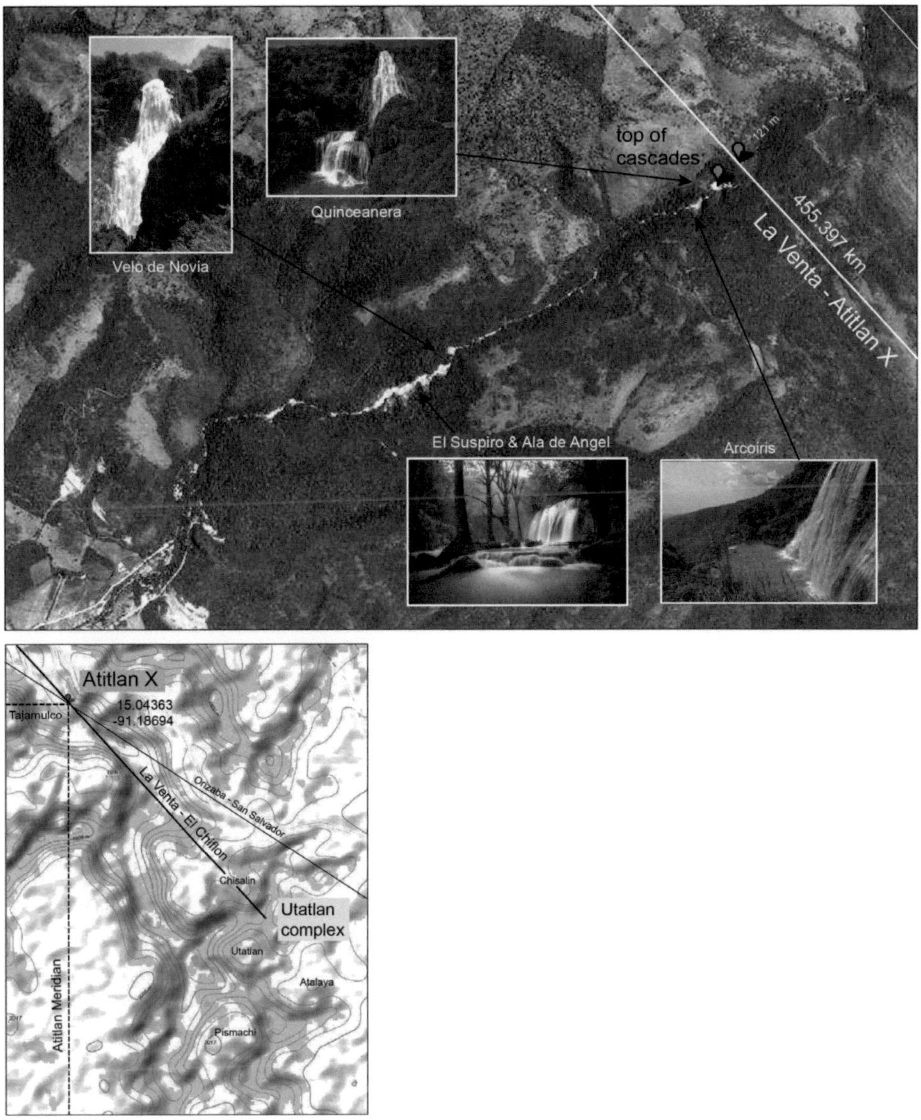


**Figure 13.**
*Satellite image of total cascades at El Chiflón, Chiapas; the proximity of large-scale line La Venta – Atitlan X to top of cascades; topo map of Atitlan X area showing the passage of line to the center of Hispanic period Mayan site of Utatlán near village where Popol Vuh creation myth was recovered (open access photos).*

Before leaving El Mirador, the archaeoastronomy work of Sprajc et al. [42] provides an opportunity to distinguish prehistoric, ritually used urban design as "space" or "time."

**Figure 14** superimposes symbolic map aspects of El Mirador over their published conclusions detailing multiple alignments of site features with astronomical phenomena toward rises in the east. Of the four superimposed lines from Danta that formally define El Mirador's major elements, only the central axis comes close to one of the astronomically significant lines. Sprajc et al.'s measurement of the azimuth from Monos to 1A 5–1 is part of their 12° group, corresponding to known astronomical alignment preferences across Mesoamerica. Their work does not associate this "time"

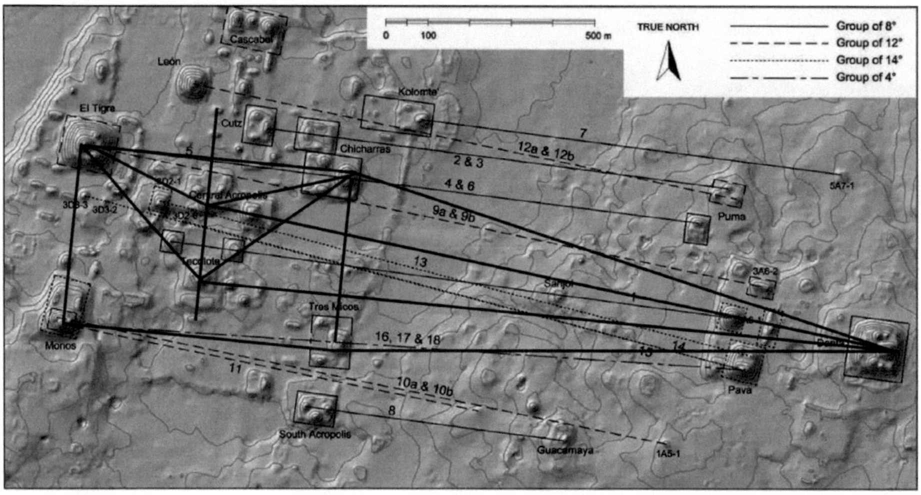


**Figure 14.**
*Map of archaeoastronomical analysis at El Mirador by Šprajc team [42:93]; present homologs superimposed in bold.*

to the Danta-El Mirador central axis as presently defined, perhaps reflecting the degree and a half difference between their Monos azimuth of 258.2334° and that of the central axis and San Lorenzo – Danta Medio of 259.76112°. The Šprajc team clearly recognizes that other factors besides astronomy may lay behind formal geometries in sites such as El Mirador:

> *"The fact that several buildings of El Mirador do not face in directions we have assumed to have been astronomically functional agrees with analogies from central Mexico and southeastern Campeche, which allow us to conclude that the placement of the main façade access to a building does not indicate the direction in which its astronomical orientation was functional, but was rather dictated by the symbolism and ritual associated with a particular structure, as well as by general factors of urban planning" [42]*

Thus, while most symbolic alignments among Mesoamerican sites and most certainly those few among the larger visible landscape studied by scholars have been motivated by studies of ritual time, these do not inform us about other, possibly more complex spatial formalities at all scales serving to powerfully organize ritual movement and social participation—once the *time* was right.

## 6. Tikal's reformation of El Mirador

Tikal's largest feature, Temple IV [47, 48], lies only about 65 km from El Mirador's Danta, possibly located by two cardinal lines, one the meridian from the point on the equinox base where Orizaba – La Venta intersects (Danta Medio), and two, a direct cardinal east from San Lorenzo. Not dissimilarly, the primary Tikal benchmark (Temple IV) involves both kinds of lines, as shown in **Figure 15**. Intent on replicating Olmec meridian power (San Lorenzo and La Venta), priests/surveyors may have prolonged the small-scale meridian from Atitlán to Atitlan X up to the east

**Figure 15.**
*Positioning of Temple IV at Tikal vertex of cardinal right angle from Santa Ana volcano meridian and point where the Atitlan/Toliman meridian strikes the Orizaba – La Venta – Danta Medio line, "Atitlan Zenith"; secondary right angle created on the line from Atitlan Zenith to Santa Ana, vertex "East X" connects rays to San Lorenzo and Temple IV.*

La Venta ray from Orizaba. This intersects "Atitlan Zenith" (10 kilometers from the Piedras Negras intersection benchmark) and becomes the latitude for a cardinal east to position Tikal. This construct clearly involved El Mirador, given that the Atitlan Zenith point aligns with the Orizaba – La Venta east ray that founded La Danta *via* the Danta Medio point on the equinox base. Temple IV's longitude on the cardinal from Atitlan Zenith appears to have been designed as the first Olmec or Mayan ceremonial site to be positioned on its own meridian directly to a volcano, Santa Ana. The line from the volcano's center runs 372.674 km precisely to the front steps of the largest of Tikal's temples (359.99814°). Tikal may also have established an association with San Lorenzo. A point (here called "East X") on the direct azimuth between Tikal's two origination points, Atitlan Zenith and Santa Ana, is an accurate right angle between Temple IV and San Lorenzo. Once one carefully orients Tikal's enormous site plan to true north with satellite images, **Figure 16**, the two locating cardinal lines do not intersect at the Temple IV point, only at the Santa Ana meridian. The latitudinal cardinal from Atitlan Zenith points instead to the northernmost Tikal called "Complex

**Figure 16.**
*Satellite photos show means of orienting published Tikal site plan; homolog azimuths between principal features.*

P." Given the substantial topography of Tikal, adjustments may have been made. The precise intersection of the two founding cardinal lines occurs in the northwest and at an elevation with limited visibility to other key site features.

Ultimately, the first line in **Figure 16** could have been a precise internal north-south meridian from Complex P prolonged down through a central feature, the East Plaza ball court, from there the homolog azimuth La Venta – Tikal was laid out as the main site axis to Temple IV. Its east extension may have been determined by running the homolog azimuth of Atitlan Medio-Tikal surveyed down from Complex P. Once this triangle symbolizing La Venta (Temple IV) and Tikal at the apex of the Santa Ana meridian (Complex P), along with the ball court vertex that connects them, was laid out, did designers go beyond the appropriation of individual homolog azimuths—designing an embedded map-like Piedras Negras, Palenque, especially El Mirador?

Opposed to the Atitlan Medio-Tikal azimuth as triangle ray is an eastern line from Tikal's Complex P whose azimuth emulates that from Atitlan Zenith directly down to Santa Ana. The azimuth from this southeast Tikal point, Temple of the Inscriptions, to the ball court does not yet map any geographical area with the Santa Ana volcano, yet individually emulates the formative El Mirador triangle ray, La Venta – El Chiflon – Atitlan X. It is true, considering only the Temple of the Inscriptions and the meridian to the ball court, that these two from Complex P are map-like, the meridian to the center as an expression of Atitlan Zenith.

The third azimuth from the Temple of Inscriptions, however, offers a clue toward a more complete homolog map. Visualize the position of the possible benchmark point of "East X," the right-angle vertex between San Lorenzo and Tikal accurately on the line from Atitlan Zenith to Santa Ana—the long east line of the partial large-scale map. Positioning the south ray of this right angle, San Lorenzo – East X, at Tikal's southeast aligns with one of the earliest ritual foci at Tikal, Mundo Perdio [49, 50]. The modest, original E-group solstice observation platform, **Figures 17** and **18**, organizes the largest Preclassic ceremonial group at Tikal. The point presently used is the midpoint of the linear feature directly east of the observer's platform of the Mundo Perdido compound. From Laporte's [50] drawings, however, astronomical observations from a western point to the rising sun to the east may have been a later addition to a location powerful for other reasons. No articulation exists, that is, markers for rise times, on the linear "horizon" feature—at least as drawn by Laporte.

Could this original feature, especially together with articulation over time, express the cascade at El Chiflón? Does this point homologically replicate El Mirador's inverted triangle apex? The azimuth from the platform point to Temple IV is La Venta – El Chiflon – Atitlan X, while its opposite ray to the Tikal center is

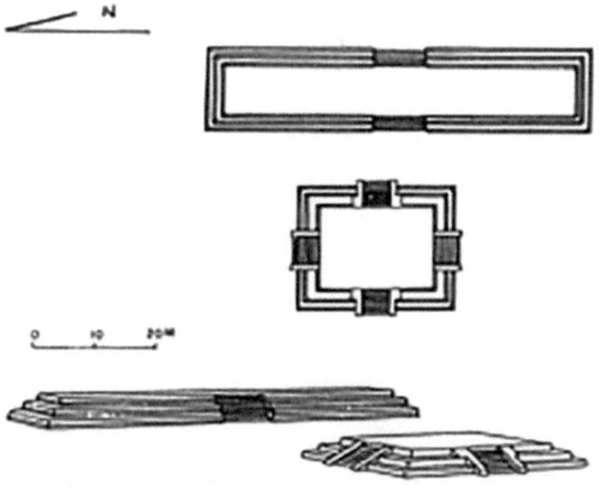


**Figure 17.**
*Linear platform focus east toward sunrise phenomena from the observation platform at the early founding E-Group of Mondo Perdido at Tikal (**Figure** 7, [50]).*

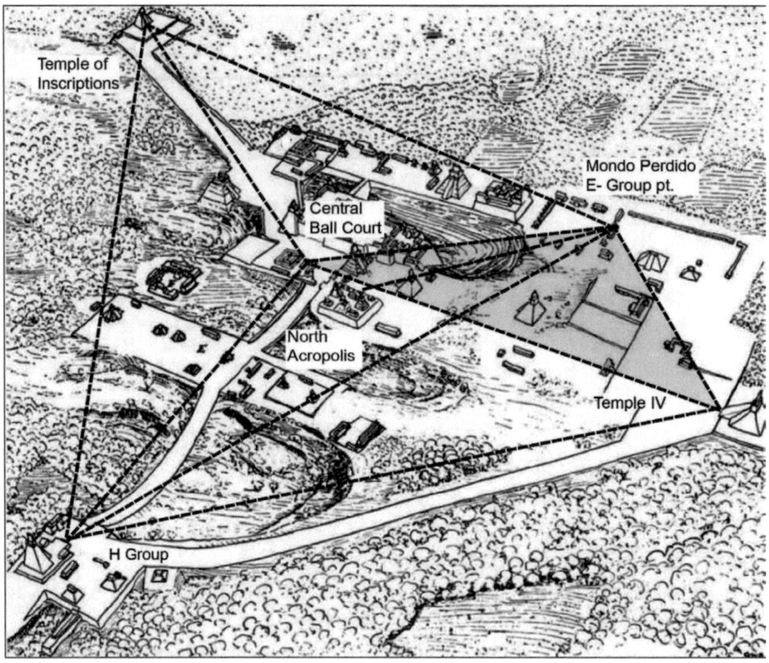


**Figure 18.**
*Perspective from Penn Museum Bulletin. The Tikal project, Vol. XX Number 4, 1956 (lines, shading, and labels added).*

El Chiflon – Tikal. Again, the azimuth from this Tikal ball court center point to the southeast is La Venta – El Chiflon – Atitlan X – parallel to the west ray of Mundo Perdido's azimuth to Temple IX. Persuasive here is this small-scale homolog's use of Tikal's major axis, the azimuth La Venta – Tikal as the third element in the Mundo Perdido triangle.

While homologically, the inverted pyramid apex point at the Mondo Perdido platform is the El Chiflón cascade, designers may have overlaid an interior axis from this point to the North Acropolis, the most impressive temple on the central axis from Temple IV to the center, **Figure 19**. Just as at El Mirador, the west triangle azimuth does not stop at El Chiflón but terminates in the powerful integrative location of Atitlan X, **Figure 20**. Associations of Atitlan X with the inverted apex may have occurred at El Mirador but are most clearly seen in the homolog map at Palenque. At Tikal, the azimuth from Mondo Perdido to the North Acropolis is Atitlan X – East X. The axis of the North Acropolis may orient to Compound P using an azimuth of East X – Danta, suggesting some greater symbolism between Tikal's most northern complex and directionally related El Mirador.

The direct azimuth from the Mondo Perdido point to Compound P may have represented this inverted apex as Atitlan X as it replicates the azimuth to Atitlan Medio (La Venta's east vertex on the equinox baseline). The coincidental symbolism of this azimuth to that from East X to Tikal (333.09274 and 332.96035), respectively, may not have been lost on priest surveyors emphasizing Atitlan X and the right-angle East X power within the Mondo Perdido homolog. This dual axis forms a good perpendicular to the azimuth from the Temple of Inscriptions to the Mondo Perdido point.

**Figure 19.**
*Tikal North Acropolis [48].*

## 7. Summary of homologic landscape pyramid mapping at ceremonial foci

Given the probable large-scale design of the Olmec landscape pyramids founding La Venta and San Lorenzo—and their symbolic unification—what proof lies in "close" mapping of large-scale azimuths in layouts of major ceremonial sites? Despite using unique, custom mapping software, presently illustrated homologs are only as good as the accuracies of published small-scale architectural and urban drawings. However,

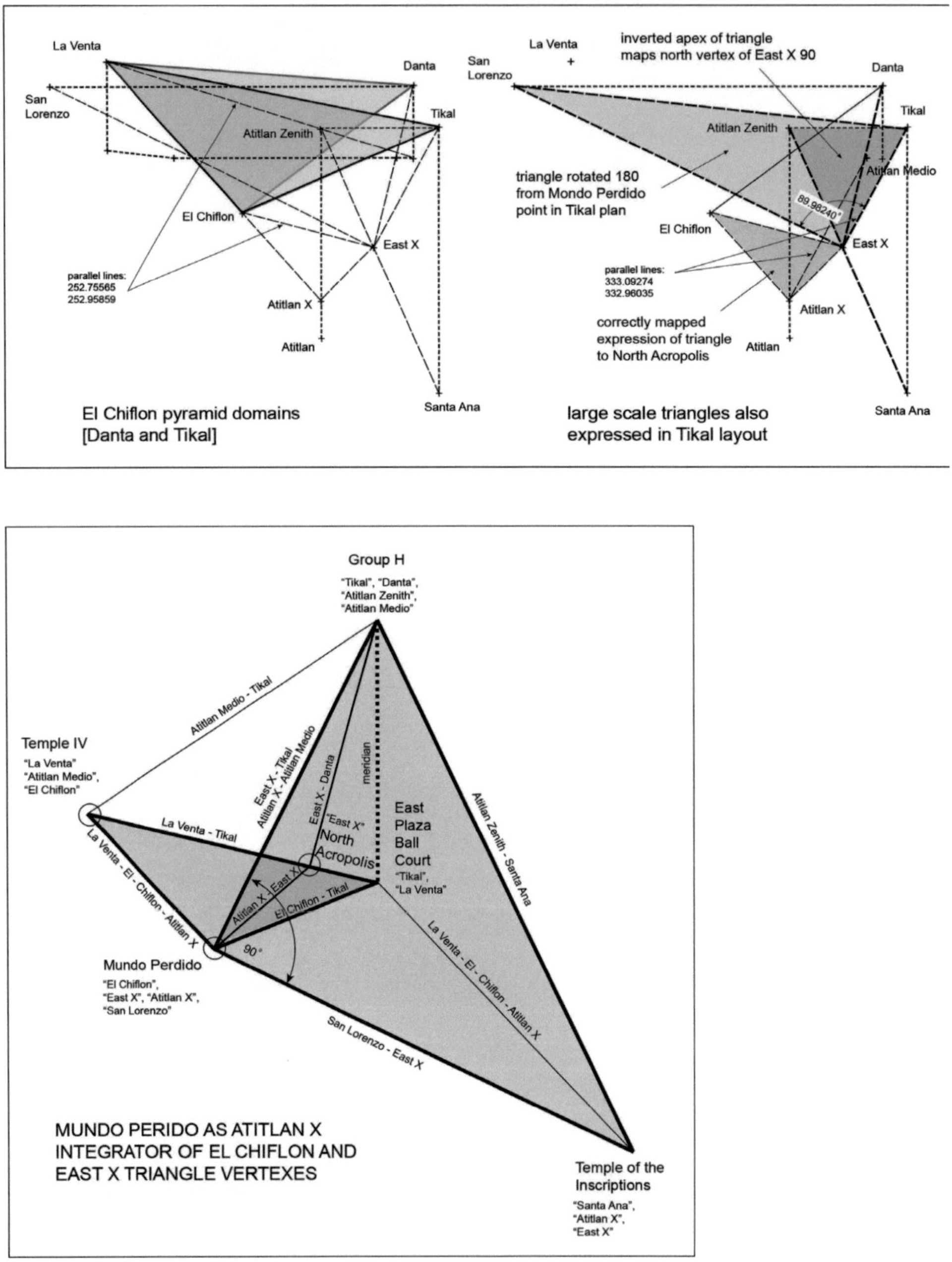


**Figure 20.**
*Principal homolog of large-scale landscape pyramid with El Chiflón at inverted vertex expressed at Tikal with Mundo Perdio E-Group as vertex, interior axis to North Acropolis is Atitlan X – East X azimuth like Palenque; Tikal's eastern triangle lines to Temple of Inscriptions are homologs expressing layout of East X in large-scale map, mediated by center axis from central ball court; other homolog azimuths are also likely.*

new archaeological mapping methodologies exist not only for lidar-imaged larger landscapes but also for precise documentation of exposed built form. What research goal will merge these not inexpensive applications, though pinning down a modest number of very accurate architectural points may not necessarily require unusual

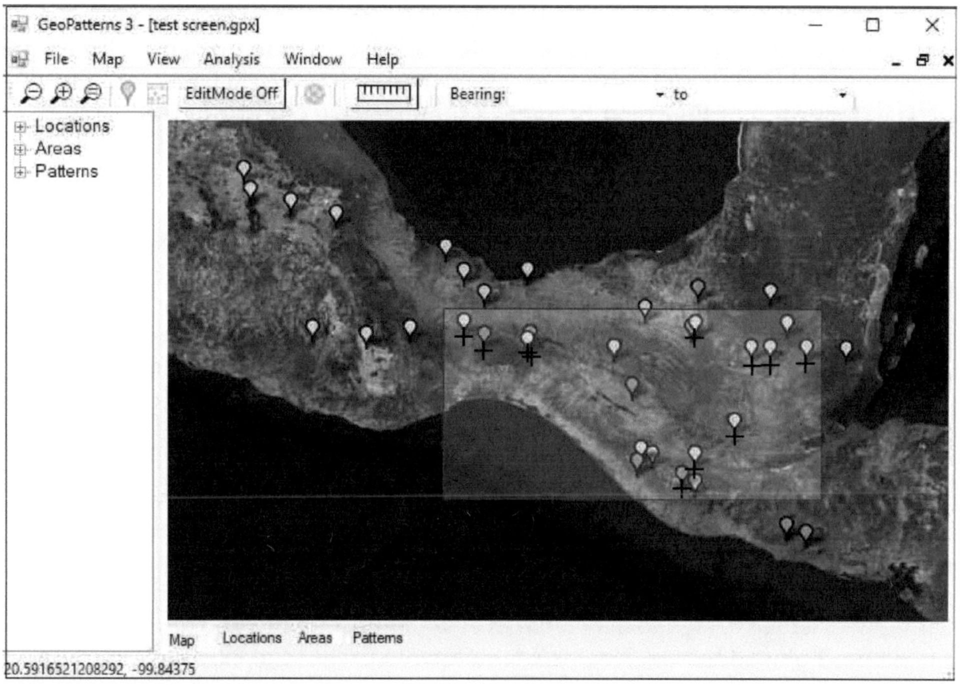


**La Venta –**
Bisect angles to Cerro del Aguila and San Salvador Medio
Meridian from La Venta Medio
Piedras Negras – Atitlan Medio

**San Lorenzo –**
Bisect angles to Cerro del Aguila and Atitlan Medio
Meridian from San Lorenzo Medio
Cerro Gordo – La Malinche – Laguna Medio – San Lorenzo Medio – Tacana
Laguna de los Cerros – San Lorenzo – La Venta Medio
Piedras Negras – San Salvador Medio

**Laguna de los Cerros –**
San Lorenzo – La Venta Medio
Meridian from Laguna Medio

**Aguada Fenix –**
Zempoaltepc – Laguna Medio
90° perpendicular to San Lorenzo – Piedras Negras – San Salvador Medio

**Piedras Negras –**
La Venta – Atitlan Medio
San Lorenzo -San Salvador Medio
90° vertex to Tres Zapotes and Piedras Negras
Equal ray distances with Monte Alban to La Venta apex
Equal vertex angles with Monte Alban to revised La Venta base point
90° vertex to revised La Venta base point and Atitlan X
Cerro Aguila – La Venta (map)

**Palenque –**
Meridian to Izapa
Bisect angles to original La Venta Medio and Atitlan Medio
Equal angles from vertices La Venta Medio and Atitlan Medio to Tzontehuitz
Atitlan Medio – San Salvador (map)
Tacana – San Salvador (map)
Tacana – Atitlan Medio (map)
Atitlan X – Atitlan Medio (map)

**El Mirador –**
Orizaba – La Venta – Danta Medio
Meridian from Danta Medio
Equinox from San Lorenzo
San Lorenzo Danta Medio (map)
La Venta – Danta (map)
La Venta – El Chiflon – Atitlan X (map)
El Chiflon – Danta (map)
Tacana – San Salvador (map)
La Venta Medio – Danta (map)

**Tikal –**
Orizaba - La Venta – Atitlan Zenith – Danta Medio
Equinox from Atitlan Zenith
Meridian from Santa Ana volcano
90° with San Lorenzo to vertex East X (aligned with Atitlan Zenith and Santa Ana)
La Venta – Tikal (map)
La Venta – El Chiflon (map)
El Chiflon – Tikal (map)
Atitlan X – East X (map)
Atitlan X – Atitlan Medio (map)

**Figure 21.**
*Test adding N, A, and C patterns to those existing among 25 archaeological and natural feature sites: comparison between adding 11 hypothetical design "benchmarks" (crosses in illustration) with 11 substituted random points within test box (above); list of patterns associated with eight Olmec and Mayan ceremonial sites (below).*

funding? To garner support, researchers of large-scale cultural geography would have to clearly state the theoretical importance of proving that large-scale azimuths were recorded, often used in locating a major site and that together with known landscape pyramid patterns, were powerful components, especially in earlier pre-textual (Preclassic) predominantly ritual organization. First, however, one needs to propose a more sophisticated means of proving design in complex geometric patterns—building on present exploratory examples for early Olmec.

The geometry of highly complex Mayan sites can be correctly assumed to have been designed, though few logical concepts are yet understood—despite the archeoastronomy of selected architectural and urban features. Likely, computers could today compute probabilities that small-scale site patterns among well-measured features are designed homologs with precisely located features in large-scale geography. The simple test of **Figure 21** provides a summarizing glimpse of future research direction. Presently, one can at least compare the effect of 11 theoretically located "benchmarks," to generate N's, A's, and C's, compared to an equal number of substituted points located randomly in the test area shown.

The existing 25 archaeological or natural feature points are first tested. They create 7 N's (nineties), 2 A's (three-point alignments), and 1 C (cardinal, either north-south or east-west); the accuracy is set at 0.06° for all three (five N's, both A's and the C fall below 0.03°). While this work early on discovered the very high (counterintuitive) number of B (bisects) created by a modest number of site points—here 25 produces 492 bisects among four points at 0.06°—right angles generally occur in numbers not unlike A's. The higher number of N's in the present case is not presently understandable.

On the software screen, one draws 11 test boxes, one over another; when turned on, each will locate a different random point. Turning off all 11 hypothetical benchmarks now generates sets with 25 existing and 11 random points each. In 10 sets, the average number of N's (existing 7 plus added random) is **13.7** compared to the total existing with added designed benchmarks of **14**; almost identical. The average total A's (existing 2 plus added random) is **7.4**, compared to the total existing including benchmarks of **34**. Few randomly produced C's create additional patterns; the total average is only **1.3**, compared to the total existing 1 plus benchmarks or **13**.

Euphoria about the comparatively very high A's and C's must be tempered by the fact that a professional architect had positioned these benchmarks, not unlike design processes creating formal geometric patterns in architectural layouts. These processes are remarkably more trial and error and time-consuming than people tend to believe. The same holds when reversing the process of design discovery. The famous computer puzzle of designing the best delivery route between a smaller number of stores across the United States—it could not be done—testifies to the complexity of design, even involving small numbers of features. Proof of design formality has always just been what is obvious to the disciplinary eye, particularly when considering images of built form. It remains to be seen, however, which readers will more intuitively appreciate the spatial logic of much more highly abstract spatial patterns defined among and in Olmec and early Mayan sites—while others will wait for statistical testing to develop.

Ethnographically, one can document how small-scale ritual practice worked to serve social purposes. Yet far less research explains the prehistoric experience of pilgrims traveling to centers like Chaco or Stonehenge—and, by extension, to Olmec and early Mayan ceremonial foci. While archaeologists add fascinating glimpses of ritual practice at several world-class sites, most notably Stonehenge, no theory yet explains how the site referenced large-scale formal patterns integrated ritual practices of distant pilgrim homelands. Particularly absent is any major shift in the *writ large* assumptions of many cultural historians that formal monumental architecture alone provides the best evidence of a "civilization" that flows one way from the center.

Reversing this conventional thinking, one can explain the location and orientation of principal Minoan "palaces" not as centrifugal, hegemonic foci controlled by great men but ceremonial religious sites integrated with others via a formal framework laid out among the most sacred mountains and caves of Crete [20]. The traditional assumption held that such foci first emerged territorially and then secondarily aligned architectural and urban features with landscape elements ("extension"); alternatively, after Minoans first designed large-scale formal landscape patterns, they secondarily located monumental architecture and its orientational features ("intension").

At Chaco Canyon, a setting familiar to the authors [13, 17, 18, 43, 44, 51], this potential conceptual shift remains on the interpretative drawing board. Park staff leads visitors to the dozen twelfth-century Great Houses and Great Kivas in northwest New Mexico, whose multistory walls still dominate this surprisingly modest canyon location. The interpretation of building technology and similarities of possible kiva rituals to extant Pueblo societies follows. Few visitors will leave the canyon thinking that this monumentality was spiritually subordinate to a totally natural appearing amphitheater wall standing just north between the two largest great houses—nor that its acoustic center point accurately aligns with the precise meridian between the two great houses on the north and south rims that first welcomed pilgrims. Despite the archaeologist, Loose's [52] careful mapping of the acoustic reach of the Curved Rock that Speaks south across the canyon to the isolated largest of great kivas, Casa Rinconada, and his calculation of huge excavation necessary to enhance the amphitheater, the rim meridian defining the ritual pilgrimage climax at the canyon center, is not yet archaeologically recognized as a *homolog* with large-scale meridians that structured the Ancestral Pueblo plateau—especially two running south from Mount Wilson and Lizard Head volcanic feature that positioned early and climax locations in the canyon [18].

Yet one will never know the ritual practiced at the great amphitheater nor how its hypothetical egalitarian effect moderated social organization among even linguistically diverse tribes on the plateau. However, linking large-scale landscape formalities to specific foci like Chaco's amphitheater begins to demonstrate that the religious power of landscape embedded concepts founded and maintained the "Chaco Phenomenon," leaving behind ideas of great houses as hegemonic territorial signs of kinship power. New methodologies of formal, symbolic spatial definition and analysis may precipitate an eventual shift in the way one conceptualizes cultural histories of the dialectic of built form and ancient landscapes.

## Author details

Dennis Doxtater
College of Architecture, Urban Planning, and Landscape Architecture, Tucson, Arizona, USA

*Address all correspondence to: doxtater@arizona.edu

## References


[1] Wilbert J. Mystic Endowment: Religious Ethnography of the Warao Indians. Cambridge: Harvard University Center for the Study of World Religions; 1993

[2] Gossen G. Temporal and spatial equivalents in Chamula ritual symbolism. In: Lessa WA, Vogt EZ, editors. Reader in Comparative Religion. N.Y: Harper & Row, Publishers; 1958. pp. 135-148

[3] Carlsen R. Footpath of the dawn, footpath of the Sun: Maya worldview at Lake Atitlán. In: Cecil LG, Pugh TW, editors. Maya Worldviews at Conquest. Boulder: University Press of Colorado; 2009. pp. 298-314

[4] Stanzione V. Rituals of Sacrifice: Walking the Face of the Earth on the Sacred Path of the Sun, A Journey through the Tz'utujil Maya World of Santiago Atitlán. Albuquerque: University of New Mexico Press; 2003

[5] Toohey J. Feeding the mountains: Sacred landscapes, mountain worship, and sacrafice in the Maya and Inca worlds. Reviews in Anthropology. 2013;**42**:161-178

[6] Wilson R. Maya Resurgence in Guatemala: O'eqohi' Experience. Norman: University of Oklahoma Press; 1995

[7] Šprajc I. Astronomical alignments at Teotihuacan, Mexico. Latin American Antiquity. 2001;**1**(4):403-415

[8] Doxtater D. Designing the "Olmec frame": Sacred large-scale surveying of Mesoamerican volcanos and pyramid shaped landscape domains; positioning of site features and their location. Unpublished manuscript. 2019

[9] Iwaniszewski S. Archaeology and archeoastronomy of Mount Tlaloc, Mexico: A reconsideration. Latin American Antiquity. 1994;**5**(2):158-176

[10] Clark J, Colman A. Olmec things and identity: A reassessment of offerings and burials at La Venta, Tabasco. Archaeological Papers of the American Anthropological Association. 2014;**23**(1):14-37

[11] Lekson S. The Chaco Meridian. Walnut Creek, California: Altamira Press; 1999

[12] Lewis MJT. Surveying Instruments of Greece and Rome. Cambridge, UK: Cambridge University; 2001

[13] Doxtater D. A report on geopatterns software: Describing and analyzing large-scale geometry between Anasazi and natural sites in the SW U.S. In: Proceedings of the Computer Applications and Quantitative Methods in Archaeology 2006 Conference. Budapest: Archaeolingua; 2008

[14] Doxtater D. Close mapping of St. Olav's pilgrimage path through Gudbrandsdal Norway: Probabilities of a designed, land surveyed concept of a large-scale Christianized landscape. International Journal of Religious Tourism and Pilgrimage. 2023;**11**(1):83-117

[15] Doxtater D. Pilgrimage to the "Vi" Center of Bulverket's water square: Mapping the location and orientation of Gotland churches. In: Martinho VJPD, da Rocha Nunes JAG, de Jesus Pato ML, de Matos Castilho LA, editors. Experiences, Advantages, and Economic Dimensions of Pilgrimage

Routes. Hershey, Pennsylvania, USA; 2023. pp. 143-190

[16] Doxtater D. Land surveying in early medieval Norway: A St. Olav pilgrimage path as a means of creating an integrated Christian society in a Viking landscape? Landscape Research. 2022;**48**(3):427-452

[17] Doxtater D. Probabilities of designed locations of ceremonial foci: The Chaco meridian, temple IV at Tikal, and a large-scale sacred Adena river landscape. Time and Mind. 2021;**14-2**:283-326

[18] Doxtater D. Chaco's place in a formalized landscape: Design and surveying of large-scale geometries as ritual practice. 2020. Unpublished manuscript

[19] Doxtater D. Probabilities of designed alignments among largest prehistoric mounds and most prominent natural features in Scandinavia – With a comparison to patterns among train stations, campgrounds and feed stores 2013; Unpublished paper available on CAPLA (college) website, University of Arizona

[20] Doxtater D. Rethinking the sacred landscape: Minoan palaces in a georitual framework of natural features on Crete. Landscape Journal. 2009;**28**(1):1-21

[21] Inomata T, Fernandez-Diaz JC, Triadan D, Mollinedo MG, Pinzón F, Hernández MG, et al. Origins and spread of formal ceremonial complexes in the Olmec and Maya regions revealed by airborne lidar. Nature Human Behavior. 2021;**5**:1487-1501

[22] Newman H. Lost Olmec City of Laguna de los Cerros in Ancient Mexico. YouTube: Megalithmania. Available from: @MegalithomaniaUK [Accessed: 2020]

[23] Rostain S, Dorison A, de Saulieu G, Prümers H, Le Pennec J, Mejía F, et al. Two thousand years of garden urbanism in the upper Amazon. Science. 2024;**383**:183-189

[24] Hillier B, Leaman A, Stansall P, Bedford M. Space syntax. Environment and Planning B. 1976;**3**:147-185

[25] Ferguson TJ. Historic Zuni Architecture and Society-An Archaeological Application of Space Syntax. Tucson, Arizona, USA: Anthropological Papers of the University of Arizona Number 60; 1996

[26] Morton S, Peuramaki-Brown M, Dawson P, Seibert J. Civic and household community relationships at Teotihuacan, Mexico: A space syntax approach. Cambridge Archaeological Journal. 2012;**22**(03):387-400

[27] Doxtater D. What visitors "do" in recreational landscapes: Using categories of affordances for evaluation, design and simulation. In: Gimblett R, Hans S-P, Muhar A, editors. Monitoring, Simulation, and Management of Visitor Landscapes. Tucson, Arizona, USA: University of Arizona Press; 2007. pp. 13-37

[28] Doxtater D. Architecture, ritual practice and co-determination in the Swedish office. In: Ethnoscapes: Current Challenges in the Environmental Social Sciences. Aldershot, UK: Avebury; 1994

[29] Doxtater D. Socio-political change and symbolic space in Norwegian farm culture after the reformation. In: Turan M, editor. On Vernacular Architecture: Paradigms of Environmental Response. Ethnoscapes Series. Vol. 4. London: Gower; 1990

[30] Doxtater D. Meaning of the workplace: Using ideas of ritual space in design. In: Gagliardi P, editor. Symbols

and Artifacts. New York, London: Routledge; 1990. pp. 107-127

[31] Doxtater D. Spatial opposition in non-discursive expression: Architecture as a ritual process. Canadian Journal of Anthropology. 1984;**4**(1):1-17

[32] Doxtater D. The symbolism, structure and politics of "center" in the old scandinavian farm culture [dissertation]. Ann Arbor: University of Michigan; 1981

[33] Hirth K, Cyphers A, Cobean R, De León J, Glascock M. Early Olmec obsidian trade and economic organization at San Lorenzo. Journal of Archaeological Science. 2013;**40**:2784-2798

[34] Malmström V. A reconstruction of the chronology of Mesoamerican calendrical systems. JHA. 1978;**ix**:105-116. (NASA Astrophysics Data Systems)

[35] Ashmore W. Spatial orders in Maya civic plans. Latin American Antiquity. 2002;**13**(2):201-215

[36] Nelson Z. Settlement and population at Piedras Negras, Guatemala [dissertation]. University Park, Pennsylvania: Pennsylvania State University; 2005

[37] Satterthwaite L, Butler M, Mason J. Piedras Negras archaeology, 1931-1939. In: Weeks J, Hill J, Golden C, editors. Piedras Negras Preliminary Papers, Piedras Negras Archaeology: Architecture. Philadelphia: University of Pennsylvania Museum of Archaeology and Anthropology; 2005

[38] Hartung H. Alignments in architecture and sculpture of Maya Centers: Notes on Piedras Negras, Copán and Chichén Itzá. Ibero-amerikanisches Archiv, Neue Folge. 1984;**10**(2):223-240

[39] Bassie-Sweet K. Maya Sacred Geography and the Creator Deities. Norman: University of Oklahoma Press; 2008

[40] Schele L, Freidel D. A Forest of Kings: The Untold Story of the Ancient Maya. New York: William Morrow; 1990

[41] Hansen R. Continuity and disjunction: The pre-classic antecedents of classic Maya architecture. In: Houston S, editor. Function and Meaning in Classic Maya Architecture. Washington, D.C.: Dumbarton Oaks Research Library and Collection; 1998. pp. 49-122

[42] Šprajc I, Morales-Aguilar C, Hansen R. Early Maya astronomy and urban planning at El Mirador, Peten, Guatemala. Anthropological Notebooks. 2009;**15**(3):79-101

[43] Doxtater D. A hypothetical layout of Chaco Canyon structures via large-scale alignments between significant natural features. Kiva. 2002;**68**(1):23-47

[44] Doxtater D. Parallel universes on the Colorado Plateau: Indications of Chacoan integration of an earlier Anasazi focus at Canyon De Chelly. The Journal of the Southwest. 2003;**45**(1 & 2):33-62

[45] Tedlock D (translator). Popol Vuh: The Definitive Edition of the Mayan Book of the Dawn of Life and the Glories of Gods and Kings, with Commentary Based on the Ancient Knowledge of the Modern Quiché Maya. New York: Simon & Schuster; 1985

[46] Carmack R, Weeks J. The archaeology and ethnohistory of Utatlan: A conjunctive approach. American Antiquity. 1981;**46**(2):323-341

[47] Ponciano E. El colosal Templo IV, Tikal, Petén Guatemala: Investigaciones,

arqueológicas y restauración en 2010. Antropologia e historia de Guatemala. 2011;**10**:131-154

[48] Loten H. Additions and Alterations: A Commentary on the Architecture of the North Acropolis, Tikal, Guatemala. Philadelphia: University of Museum of Archaeology and Anthropology Tikal; 2007. Report 34, part A 2007, fig. 49

[49] Almers J, Rice P. Astronomy, ritual, and the interpretation of Maya "E-group" architectural assemblages. Ancient Mesoamerica. 2006;**17**:79-96

[50] Laporte J, Fialko V. Un Reencuentro con Mundo Perdido, Tikal, Guatemala. Ancient Mesoamerica. 1995;**6**(1):41-94

[51] Doxtater D. Reflections of the Anasazi cosmos. In: Gron O, Engelstad E, Lindblom I, editors. Social Space: Human Spatial Behavior in Dwellings and Settlements. Odense: Odense University Press; 1991. pp. 155-184

[52] Loose R. Tse'Biinaholts'a Yalti (curved rock that speaks). Time and Mind: The Journal of Archaeology, Consciousness and Culture. 2008;**1**(1):31-50

Chapter 6

# The Role of Elements of Small Folk Architecture in the Czech Landscape

*Vendula Moravcova, Jana Moravcova and Denisa Pekna*


## Abstract

This chapter focuses on the Czech cultural landscape, mainly created during the Baroque period. At that time, humans transformed natural settings to enhance their aesthetic appeal. This transformation involved constructing forest parks, alleys, vista axes, and landscape landmarks. Alongside grand complexes, people built small monuments of folk architecture in rural areas. Although many sizeable cultural landscape elements have been transformed or forgotten over time, small monuments of folk architecture still exist in today's landscape. This work aims to map and evaluate the aesthetic qualities of these small monuments of folk architecture in selected parts of the microregions of the Czech landscape.


**Keywords:** landscape architecture, small folk architecture, history, landscape aesthetics, Czech cultural landscape

## 1. Introduction

Around the eighteen centuries, the phrase folk architecture was first used, which marked the first developmental stage of building architecture. This stage mainly contained simple construction elements and procedures used in folk construction [1]. According to some authors, the term folk architecture began to be used after that, lacking a specific sense of meaning. Architecture is understood as monumental buildings built by a professional architect, which is missing from folk construction. For this reason, in the second half of the twentieth century, some authors began to promote new terms such as anonymous architecture, spontaneous architecture or architecture without architects. However, these new concepts were not experienced, so we still use folk architecture today [2]. The very concept of folk architecture is understood differently today. It is often used to express the historical heritage of various residential or economic buildings [3]. Sometimes, this term is used in connection with technical buildings such as forges and mills. Many authors of folk architecture also include small sacral objects, i.e., chapels, divine torment and others [4].

Small folk architecture is commonly used for small sacral objects, border stones, memorials or old signposts. The monuments of folk activity are noticeable in the landscape to this day. Unfortunately, many of them are damaged by time, natural influences, and human activities, and some of them have become the aim of theft and organised crime with cultural and historical monuments. Repair of the elements of small folk architecture and their return to the landscape care of voluntary associations and The Ministry of Culture and the Ministry of Agriculture [5].

## 2. History of folk architecture

### 2.1 Prehistory and antiquity

The beginnings of man's construction activity cannot be precisely determined in time. The reason for these activities was to ensure the basic need for safety and protection from adverse natural influences. At first, simple screens and primitive shelters were used for this. These first buildings were then replaced by the first huts, the walls of which enclosed the defined area space. More complex buildings were later made of stone foundations, clay walls and a wooden roof. Bricks also gradually began to appear to have dried in the sun [6]. Thanks to bricks, the first more complex types began to appear in prehistoric times, such as walls, pillars, lintels and vaults. They were first built in the territory of Europe, large, up to 10 metres long, houses. These were made up of wooden support poles, between which wicker was woven [7]. The entire building was then protected from above by a simple one-gable roof. Later, the central courtyard adjoined the rectangular residential building, where the family spent the most time. This space was also used to light and ventilate the rooms [8]. In these times, the first buildings began to appear, which can be defined as small folk architecture. These were primarily primitive buildings associated with pre-Christian and early Christian religion and sacred places. The constructions were primarily simple places of prayer and places of sacrifice.

### 2.2 The middle ages

Most information about the features of the original medieval village house provides archaeological research. They were focused on knowledge from the beginning of the medieval village and its original dwelling. However, they made up most of the objects found in defunct houses that left behind only barely legible remains. The most extensive organic material that was decomposed by microorganisms succumbed to destruction. Clay, which, together with wood, formed the most common building material, was subject to destruction by influence transformation [2]. If we leave aside some prehistoric and Celtic archaeological finds, the history of Czech architecture begins in the ninth century. At that time, it spread over our territory, the Great Moravian Empire, where a system of fortifications was built. Archaeological discoveries from this time prove the existence of churches and many buildings with different floor plans. In addition to the typical rectangular one-nave floor plan, a cruciform floor plan was found in typical basilicas and rotundas floor plans. The buildings were made of quarry stone joined together with mortar. Colourful frescoes were often found on the plaster inside the buildings. In the century by the eleventh, many small churches were being built in the countryside [8].

Although the Middle Ages lasted in the Czech lands until the beginning of the sixteen century and was characterised by a substantial expansion of Christianity, folk architecture can also note significant elements from pre-Christian times. These are primarily references to original folk myths and spiritual forces of nature. These references are then visible in the decorations of folk architecture, which, in contrast to the style architecture, uses a more accessible use of materials and decorations, taking into account the available raw materials and local customs [4].

### 2.3 The modern age and subsequent development

Since the High Middle Ages, folk architecture has been continuously developing. The modern development of vernacular architecture is based on general building development, which is very different. Building styles such as Renaissance and Baroque are often combined in the same building; therefore, its unequivocal classification is impossible. This fact is a given, especially the time delay between the given style and its application in the countryside [4].

At the beginning of the Renaissance, vernacular architecture thus reflects the general construction development. The elements of the style of architecture were transferred to the village buildings using manor buildings and farms. These buildings were the first to impact the style of architecture. Thanks to this, they served as inspiration for modifications of country houses. However, the Renaissance had little influence on the arrangement of the village house.

A significant turning point for small folk architecture occurred in the Baroque era. In the period after the Battle of the White Mountain (1620), which represented an important milestone in the historical development of the Czech lands, there was an era of often violent re-Catholicisation of Bohemia and the suppression of other religious currents. At this time, the construction of new shrines and places of pilgrimage flourished in the folk environment. Another significant phenomenon manifested in the rural environment was the phenomenon of the so-called Czech cultural landscape, where free nature was treated as a landscape park [8].

In combining these two phenomena, numerous avenues pass, and views of landscape landmarks and other landscape elements were built in the Czech landscape. These decorative elements in the landscape were then supplemented with elements of small folk architecture, especially of religious significance. In the Baroque era, many new ways of the cross, torments of God, crosses, chapels and other elements of small folk architecture appeared.

An extraordinary phenomenon of Czech folk architecture is building so-called reconciliation crosses. These stone crosses can still be found at crime scenes as evidence of the perpetrator's repentance and an eternal reminder of the event.

The phenomena of Baroque architecture remained in people's consciousness even after the following period. Although the folk environment did not change much during the following centuries, the period of the nineteenth century brought to folk architecture romantic ideas about historical housing and the outbreak of construction in the spirit of historicising baroque called Peasant baroque, which was mainly influenced by the later Baroque architecture [9]. This short stage brought new architectural elements into the folk environment through bright colours and distinctive stucco decorations. Buildings are mainly typical for decorating shields and gates of village buildings. These decorations were created in the original baroque style, formed by an unusual combination of various decorative elements [10]. The rounded silhouettes of the shields often resemble the shapes of the landscape. Some shields even contain elements with nature motifs. Thanks to the richness and diversity of these buildings, the Peasant Baroque became an architectural gem of South Bohemia [11].

Another critical period for the Czech folk environment was the era of the two world wars. The period of the wars themselves did not directly affect the people's environment, but the events following their end are still visible today.

The period after the First World War and the establishment of the independent Czechoslovak Republic meant for the public environment the construction of monuments commemorating war victims, especially those who did not return from the battles at the front, and monuments to national identity—in contrast, the period after the Second World War brought destruction to a considerable number of folk monuments. After the end of the war, there was an era of the removal of the German population from Czech cities and villages, which often meant absolute depopulation of villages, especially in the border areas. The Czech borderland, inhabited by the German population since the Middle Ages, remained almost abandoned and unprotected against attacks by the Czech population, who often tried to remove everything resembling the original German settlement. During this turbulent period, honourable churches, chapels and small monuments of folk architecture were also destroyed.

The border areas subsequently underwent further destruction after 1948, when a border area was established, complete with military areas. The buildings of the original folk architecture remained uninhabited and decayed to the point of being uninhabitable or falling victim to artillery training. The paradox of this period is that landscape elements such as chapels and crosses were preserved as landmarks in an often-confusing landscape [12].

In areas outside the border zone, folk architecture was more or less preserved, but in many cases, the rural environment was interfered with unsparingly. The cottage industry phenomenon, which is very widespread in Bohemia, dramatically influences this. The owners of the original houses, which are not under heritage protection, often rebuild the buildings unsparingly, regardless of the original character.

Currently, small monuments of folk architecture are facing new challenges. Many small monuments, especially peace crosses and statuettes from chapels and crosses, become the target of raids by antiquities dealers who export these monuments abroad as decorations. This trade in historical objects is remembered in Act No. 71/1994 Coll., on the sale and export of cultural objects values, more precisely in its Annex No. 1, "list of objects of cultural value," where the protected elements of minor architecture are listed such as bells, torments of God, field and reconciliation crosses and their fragments and parts [13].

Another threat is the frequent aimless intentional damage to seemingly insignificant remains of the past and, finally, the lack of financial resources to reconstruct a large number of dilapidated buildings [14, 15].

On the other hand, some new elements referring to folk architecture are returning to the Czech landscape. Various voluntary groups try to repair or return to the landscape damaged or disappeared elements of small folk architecture.

Nowadays, the Czech landscape is filled with new elements that, on a conceptual level, resemble the original works of folk architecture. It is a modern trend to build crosses or memorials at the sites of tragic events or severe traffic accidents, which could be compared to the ideological legacy of the original monuments of small folk architecture.

## 3. The origin and function of small sacral buildings

Living in an unknown and unstructured landscape always brings with it a feeling of uncertainty and danger. That is why people have been looking for unique places in the landscape since ancient times, and they attribute a special meaning to them. In addition to hills, wells and fords were important forks and junctions. Besides their orientation function, Crossroads symbolised choices between evil and good, i.e., the bad and the good path. People in these places, to enlist a specific supernatural power to help them in difficult decisions, planted trees and piled stones [16]. In pagan times, there were various signs or small ones in these places, such as stone pyramids or menhirs. In the beginning of Christianity, the church wanted to limit these pagan traditions and thus enrich the existing objects, such as the symbol of the cross and the body of Christ. She then placed holy images on the sacred trees. In other cases, the church destroyed pagan objects, and she built Christian buildings in their place. For this reason, the first objects were created in small folk architecture in the landscape [17].

## 4. Typology of small folk buildings

### 4.1 Crosses

Crosses are probably the oldest element of small sacral architecture. In the earlier ones, a procession sometimes went to them during mass. It was an honour for the cross owner when his cross served these religious purposes [18]. Crosses also served for orientation or to mark important points [19]. We can find them at the beginning and end of villages, near cemeteries and churches, on the border of plots or on the way to places of pilgrimage [20]. Due to the edict issued by Charlemagne, which commanded the building of crosses around roads, their construction spread mainly in the ninth century. The first crosses were made of wood. Due to workability, availability and price, wood was a very suitable building material, especially for the poorer sections of the population. Due to the material's durability, these original wooden crosses were not preserved. More durable, only rich and influential people allowed themselves to build crosses made of hewn stone [17].

As evidenced by the development in Wallachia and Slovakia, where, alongside wooden crosses, they began gradually building stone crosses with Christ painted on a tin plate. From the middle of the nineteenth century, these crosses gradually replaced cast iron crosses, which spread mainly in the areas of domestic smelters, where they were manufactured. On these crosses, we can find other cast iron elements such as the corpus of Christ, a figure of the Virgin Mary, or a decorative inscription on a tablet. They made stone plinths by local stonemasons. Various inscriptions and symbols were carved into the plinths, mostly chalices or hearts. Less expected were the engraved names and residences of those who had the cross built. Later, marble crosses were also built.

The so-called rooster cross is a particular type characteristic of the Sumava region. This particular type of cross had the task of telling the story of the crucifixion of Jesus Christ, where, in addition to the crucified body, other attributes closely related to this event are placed on the basic cross structure [21].

### 4.2 Chapels

A chapel, from the Latin word Capella, is a small building built to worship saints. They are most often found in the centre of villages, where they replace the function of a church. There are also chapels built in prominent places such as wells, hilltops, and pilgrimage routes. Chapels are also part of the Stations of the Cross, which usually have fourteen stops [13]. Smaller buildings, forming a transition between the torments of God and chapels, are niche chapels. We divided them into older niche chapels, a part of a door leading into a small niche, and younger niche chapels, which were considerably more spacious. Into niches, often protected by glass or a grid, they housed statues or religious images [22]. Most chapels come from the Baroque when their number expanded the most. The niche chapels were an advantage because of the low costs required for their construction [23].

For this reason, they were built not only by townspeople but also by people from the villages. In some places, they were also built as Stations of the Cross. In connection with the Stations of the Cross, there are niche chapels, sometimes called bows [24].

### 4.3 Bells

The first folk bells, preserved in our territory, originate from the sixteenth century. These bells were used not only to announce deaths but also to summon people for work, midday prayer or to remind them about the time to kneel. The fire patent issued in 1751 by Maria Theresa contributed to expanding the number of country bells. This patent mandated that a fire be announced by ringing a bell. Because of his non-compliance, which was issued by Joseph II. in 1787, the "fire reporting order" was restored. The bell was first suspended in the natural tree trunk, then on the so-called fork, i.e., the trunk of an old tree with a canopy, usually in the middle of the village. Later, bells were placed in village chapels and small chapels [12].

### 4.4 The torments of god

The torments of God are small stone or brick buildings. Most often, they have the shape of a column or pillar due to the space needed for placing a statuette, painting or relief in the upper part. The torments are equipped with a cross at the top. As we can already guess from the name, it is a reminder of the torture of Christ in ancient times stone column. The reasons for God's torments were different. They were built as thanks for warding off various threats and epidemics or in return for requested help in an emergency. Other reasons were honouring the dead or commemorating important family or local events [25]. According to some contemporary records, it was the torment of God around the paths to the gallows, where they allowed the condemned to say their last prayers. Some authors also state that torments in southern Bohemia are associated with war events. More precisely, it refers to the marking of the places where fallen soldiers buried the remains. God's torment here served its purpose and signified a violent death [13]. We divided the church into stone and brick according to the materials used. The stone torment of God was carved and had three parts. The first part was the pedestal, most often a square floor plan. The second part consisted of a round (torments post) or angular pill (pillar flour). The pillar or pillar also had three parts: the foot, stem and head. The last part of the stone torture chamber was the chapel, which formed the most crucial part of the whole object. Masons carved niches into the chapels, which they subsequently entered, and they placed painted images on wooden boards or sheet metal. In some cases, the depicted scenes were directly carved into the niches. He was on the upper part of the chapel, where a top stone or forged cross is usually placed [22].

### 4.5 Peace crosses

The Reconciliation Cross is often an asymmetrical cross carved from a single piece of stone, serving to straighten out the relationship between the killer and the family of the deceased. In the sixteenth and seventeenth centuries, the fate of the murderer was decided only by his closest relatives, who were with the murderer and concluded a so-called reconciliation agreement [26]. Part of this contract was not only financial compensation but also recognition of guilt and the obligation to erect a cross at the site of the murder [13]. The surface of some crosses also has a killer carved on them, as well as the subject, year, and inscription commemorating the event [27]. The basic types of these crosses include cross stone, stone cross and cruciform bike. For some areas, individual types are characteristic. In Chebsk and northern Bohemia, stone crosses predominate. In the vicinity of the cities of Pilsen, Domažlic, Tachov, Pelhřimov, Jihlava and Třebíč, we find more cross stones. Cross bikes can only be found in the Plzeň - north district. All types occur together in other parts of the country [28].

### 4.6 Ways of the cross

The Way of the Cross consists of buildings or art depicting Christ's suffering. From the fifteenth century, the Stations of the Cross consisted of a cycle of seven stops. Today's most common form of the Way of the Cross, with fourteen Stations, has been recorded since the Baroque period [11]. Originally, paintings or sculpture engravings inside temples and churches depicted the Stations of the Cross. In the Baroque period, she took the images outside and exhibited them on special occasions. It was the rise of the outdoor construction of the Stations of the Cross and its stops. These stops are often built on the hillside. The journey was more arduous after that, and it represented the Passion of Christ better [29, 30].

List of fourteen stops according to Mecchi [28]:

I. Stop: Jesus condemned

II. Stop: Jesus accepts the cross

III. Stop: Jesus falls for the first time

IV. Stop: Jesus meets the mother

V. Stop: Simon helps to carry

VI. Stop: Veronika wipes his sweat

VII. Stop: Jesus falls a second time

VIII. Stop: Jesus comforts the women

IX. Stop: Jesus falls the third time

X. Stop: Jesus stripped of his clothes

XI. Stop: Jesus nailed to the cross

XII. Stop: Jesus dying on the cross

XIII. Stop: Jesus on his mother's lap

XIV. Stop: Jesus buried

### 4.7 Other

Although these are not typical monuments of small folk architecture, finding other small historical objects closely related to life in the Czech countryside is possible. These are, for example, preserved boundary stones or historical signposts. These small monuments can most often be found at the intersection of critical historical roads and on the former borders of lands belonging to critical noble families. Border stones are usually oriented horizontally with the border of the land when the stones are marked with the signs of the families to which the given territory belongs [22, 31].

## 5. Material

Two microregions were chosen for this work: Blanicko-Otavsky region and region Straziste. Both localities have decadent natural and historical richness and are concentrated around the local city.

As representative material for both microregions, the central cities were chosen as their administrative centres. The microregion Blanicko-Otavsky is the city of Protivin, and the microregion Pacovsko is Pacov City (**Figure 1**).

### 5.1 Microregion Blanicko-Otavsky

In 1999 there was found the "Association of Municipalities of the Blanicko-Otava region", whose name was derived from the rivers that flow most of the member municipalities, which included the towns of Vodňany, Protivín and the surrounding

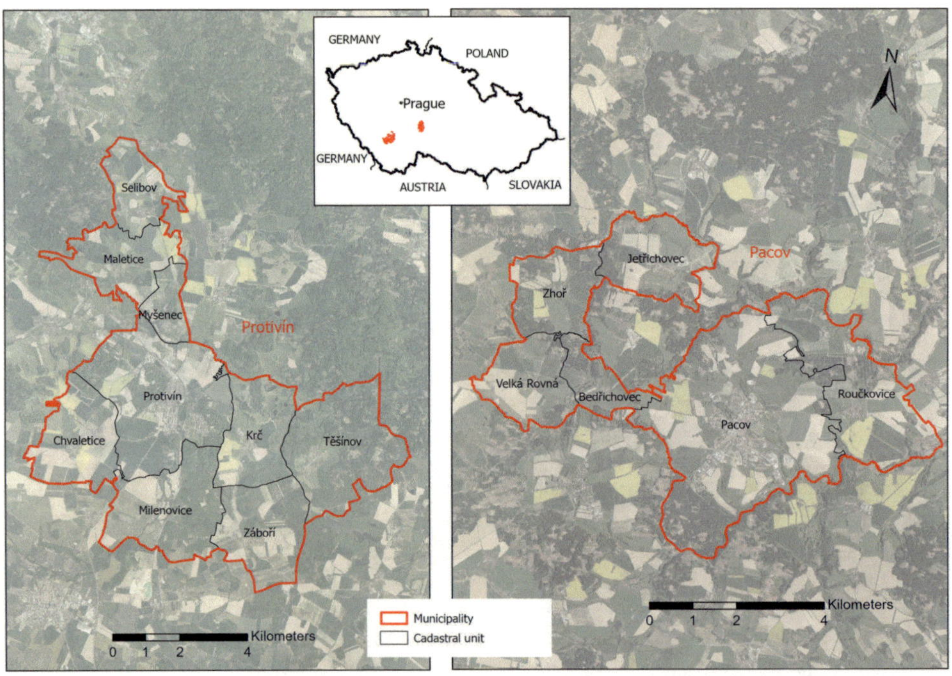


**Figure 1.**
*Localisation of microregion Straziste and Blanicko-Otvasky.*

municipalities in Vodňany. Its aim was the provision of the association, which would protect common interests and multiply the strength and means of promoting intentions, exceeding the extent and importance of the individual municipalities.

Nowadays in the Blanicko-Otavsky region, they are towns Bavorov, Protivin and Vodnany and villages Bavorov, Bilsko, Cicenice, Dolni Novosedly, Drahonice, Herman, Krajnicko, Kraslovice, Mekynec, Paseky, Pivkovice, Pohorovice-Kloub, Putim, Skaly, Skocice, Talin and Zdar.

The main centre of the microregion is the town of Protivin which is in the centre of the microregion.

### 5.2 Municipality Protivin

Protivin, named after first holder Protiva, originates around 1260 as a fortress and a village on the ford through the Blanice River, in the area of gold-bearing streams and mountains. The first official notice comes from 1282 when the kings Charles IV and Wenceslas IV used the fortress as a hunting castle. In 1562, King Ferdinand I sold the estate of Protivin to the ownership of the Lords of Hradec. At this time, the administrator of Zelendar was founded by a system of ponds northeast of Protivín. Other holders became the Vratislav brothers of Mitrovice. In 1711, Protivín became the property of the Prince of Schwarzenberg as a town, a fortress, a brewery, a mill, a saw and others. At the end of the nineteenth century, the city developed, but then due to unemployment after the closure of the sugar refinery and the wave of emigration, especially to the US, where the state Iowa was founded in 1872, the settlement, which was named Protivin.

In August 1899, Protivín was promoted to the city. Currently, Protivin has about 5000 inhabitants with connected municipalities. The city of Protivín consists of several cadastral territories; in their area, there are several surrounding villages administratively belonging to the town of Protivín. These cadastral territories are Chvaletice u Protivina, Krc u Protivina, Maletice, Milenovice, Mysenec, Protivin, Selibov, Tesinov u Protivina and Zabori u Protivina.

### 5.3 Microregion Straziste

The Stražiště microregion is located in the western part of the Vysočina Region, where it is directly adjacent to the South Bohemian and Central Bohemian regions and is named after the hill, which dominates the Pacovska highland. It was established in 2001 when the mayors of the membership municipalities signed a "statement of mayors of municipalities on the partnership". The reason for the microregion was primarily a more accessible and more efficient solution to the problems that the members of the municipalities connected. In the beginning, the microregion brought together a total of twenty municipalities, which were the town of Pacov and the towns of Lukavec, the villages of Bratrice, Cetoraz, Caslav, Dul, Es, Kamen, Lesna, Mezilesi, Posna, Salacova Lhota, Samsin, Techobuz, Utechovice pod Strazistem, Velká Chyska, Vysoka Lhota, Vyklantice, Zhorec and Zlatenka. In 2015, Dobra Voda u Pacova and Vezna connected to the microregion.

The main focus of the microregion is to provide several services for its members. One is the subsidy service, from preparing applications and project management to its implementation.

The main centre and headquarters of the microregion are the cities of Pacov, located on the border of the Vysocina Region and the South Bohemian Region, specifically between the towns of Pelhrimov and Tabor.

### 5.4 Municipality Pacov

The town of Pacov is the centre of the Microregion Straziste. The town's history dates back to the second half of the thirteenth century when the territory around the then-royal castle was gradually settled. The first written mention of Pacov is found in the book Rosenberg in 1316, where a dispute over the property of Oldrich of Pacov is described. Pacov recorded the most significant development in the Gothic period when the city received the first urban privileges—the following century meant stagnation in development. New developments emerged in the nineteenth century, including economic, cultural, and federal life development.

According to the last census of Pacov, about 4750 people live there.

The town of Pacov lies on several cadastral territories, including the surrounding villages that administratively belong to the whole of Pacov town. These cadastral territories are Bedrichov near Zhorice, Jetrichovec, Pacov, Rouckovice and Zhor near Pacov.

## 6. Methods

In the area of chosen central cities, all objects of small folk architecture were mapped. For accurately mapping local monuments of folk architecture, an off-road survey leaning on available map works was carried out:

- I. military mapping
- II. military mapping
- III. military mapping
- Basic map of the Czech Republic 1:10000
- Topographic map in the S-1952 system
- Local tourist map

After approximately determining the occurrence of elements of small folk architecture through map materials, a field survey was subsequently carried out to determine these monuments' actual location and cell state.

After map and terrain mapping, the list of small folk architecture monuments was made.

## 7. Results and discussion

### 7.1 Municipality Protivin

Elements in nine cadastral territories represent the municipality Protivin.

#### *7.1.1 Cadastral territory of Chvaletice u Protivina*

Area: 6,12 $km^2$.

Number of identified objects: 6.

*DOI: http://dx.doi.org/10.5772/intechopen.1006349*

Cadastral territory Chvaletice u Protivina includes one village, Chvaletice. In the cadastral territory, five crosses were identified near the roads. The crosses are metal on the stone base. The cross near the north road from the Chvaletice village is partly damaged with an unreadable sign on the base.

The following identified object is the neogothic chapel of St. Venceslaw.

#### *7.1.2 Cadastral territory of Krc u Protivina*

Area: 6,01 km$^2$.

Number of identified objects: 20.

Cadastral territory Krc u Protivina includes village Krc. In this cadastral territory, there are fourteen crosses around the roads, in the visible distance from one to another. In the intravillan, the crosses are made from stone with metal statues of the crucified body. The crosses outside the village are mostly metal with stone bases. In the centre of the village are the chapel of St. John Nepomuk and two monuments of victims of World War I. One is placed in the centre of the village, and the second is in the local cemetery. Outside the village are the torments of God, near the On Rejcovych pond and two border stones, which define earlier Schwarzenberger land.

#### *7.1.3 Cadastral territory of Maletice*

Area: 5,38 km$^2$.

Number of identified objects: 6.

The cadastral territory has the same name as the village which lies here. In the area, there are three crosses: in the centre of the village and the southern part, there are metal crosses on the stone bases near roads; on the northern end of the village, there is a granite cross dedicated to the village and the year 1886. In the centre of the village is a chapel and memorial to victims of World War I. On the west end of the village, there is a torment of God with the year 1783.

#### *7.1.4 Cadastral territory of Milenovice*

Area: 8,21 km$^2$.

Number of identified objects: 7.

In the cadastral territory lies a village of the same name. In this area, there are five metal crosses on the stone base around roads. In the centre of the Milenovice village are a chapel of St. John Nepomuk and a memorial to victims of World War I.

#### *7.1.5 Cadastral territory of Mysenec*

Area: 2,15 km$^2$.

Number of identified objects: 8.

In the cadastral territory, there is a village of the same name. From the identified objects, four crosses were found around the roads. They are all metal on the stone base. The village has three chapels: one on the west end, dedicated to the St. John Nepomuk, and two chapels are built on the cemetery wall. They both are damaged; one is dedicated to the St. Havel, and the second is without known dedication. In the centre of the village, there is also a memorial to the victims of World War I.

### *7.1.6 Cadastral territory of Protivin*

Area: 13,31 km$^2$.

Number of identified objects: 22.

In the cadastral territory lays the administrative centre of the town of Protivin. In the area of the cadastral territory, 11 crosses were identified, made of metal and stone. The crosses are placed around the roads visible to each other. In the centre are six chapels dedicated to St. Barbara, St. Anne, Virgin Mary, St. Wenceslaw and St. John Nepomuk. There are three memorials in the town: to the victims of world wars, to the politic Antonin Svehla, and to the opera singer Marta Krasova, born in the Protivin. Apart from these small architectural monuments, two historical signposts can be found in the town.

### *7.1.7 Cadastral territory of Selibov*

Area: 3,91 km$^2$.

Number of identified objects: 5.

In the cadastral territory, there is a village of the same name. In this area, there are three granite crosses next to the road. In the centre of the village, there is a landing chapel. Except for the chapel, there is a memorial to the victims of World War I.

### *7.1.8 Cadastral territory of Tesinov u Protivina*

Area: 9,55 km$^2$.

Number of identified objects: 9.

In the cadastral territory, there is a village named Tesinov. In this area, six crosses are lining the road. These crosses are metal on the stone base. In the centre of the village, there is a chapel dedicated to the Virgin Mary and a memorial to the victims of World War II. Near the village are memorable trees where a picture of the Virgin Mary is hung.

### *7.1.9 Cadastral territory of Zabori u Protivina*

Area: 6,73 km$^2$.

Number of identified objects: 9.

In the cadastral territory, there is a village named Zabori. Eight crosses are lining the road in this area, and the next is always visible. Except for the crosses, a chapel to St. Florian and St. John Nepomuk is in the village's centre.

## 7.2 Municipality Pacov

Elements in six cadastral territories represent the municipality of Pacov.

### *7.2.1 Cadastral territory of Bedrichov u Zhorce*

Area: 1.90 km$^2$.

Number of objects found: 3.

In the cadastral territory, there is a village named Zhorec. Although there are two crosses in this area, both crosses are made from metal with a stone base. The third building is the *square* Chapel of the Virgin Mary, built in 1935 in the centre of the village.

#### 7.2.2 Cadastral territory Jetrichovec

Area: 3.99 $km^2$.

Number of objects found: 5.

In the cadastral territory, there is a village of the same name. In the cadastral territory, there are three metal crosses with stone bases. In the centre of the village is a chapel initially dedicated to St. John of Nepomuk. In 2014, the chapel was reconstructed and consecrated again. In the centre of the village, there is also a memorial to the victims of both world wars.

#### 7.2.3 Cadastral territory of Pacov

Area: 16.12 $km^2$.

Number of objects found: 46.

In the cadastral territory, Pacov is the administrative centre of the whole town. We find many small buildings in Pacov and its surroundings, several of which are protected. In the area of cadastral territory, there are fourteen crosses, mostly of metal origin with a stone base. One of the most important historical monuments is a niche chapel before the castle Pacov. This chapel has a copy of the original statue of St. John of Nepomuk. In the area are six other chapels, the most important of which is the baroque Chapel of St. Anna, built on the alleged medicinal source. By the road leading to the village of Rouckovice, there are the stone baroque torments of God from 1678, probably the oldest in this area.

In the cadastral territory, there are many memorials, especially to the fallen and captives of the world wars, as well as influential personalities and events related to the village of Pacov, among which Czech writer Antonin Sova is very famous.

Apart from these minor monuments, the area has many statues and holy images in the surrounding area.

#### 7.2.4 Cadastral territory of Rouckovice

Area: 6.95 $km^2$.

Number of objects found: 8.

In the cadastral territory, there is a village of the same name. This area has five crosses, four around the village and one on the border with neighbouring cadastral territory. A chapel is dedicated to the St. John Nepomuk except for the crosses. In the cadastral territory, there are also two memorials to the victims of world wars.

#### 7.2.5 Cadastral territory Velka Rovna

Area: 3,47 $km^2$.

Number of objects found: 4.

In the cadastral territory, there is a village of the same name. In the area, there are three crosses and one chapel. The first cross is made from metal with a stone base and is built in the middle of the meadow, accompanied by two old trees. At the meadow north above the village stands the second cross, and the third cross is in the centre of the village. The chapel mentioned above is in the village's centre, dedicated to the holy guardian angels.

#### 7.2.6 Cadastral territory of Zhor u Pacova

Area: 3,43 $km^2$.

Number of objects found: 10.

All found objects in this cadastral area are formed by metal crosses. Their best preserved can be found at the cemetery wall in the village of Zhor. This cross has an interesting inscription: "Do not leave us, the Mother of Zhor, when we call you!". Near the second cemetery, located behind the village, there are three more crosses. Two can be found directly in front of the entrance gate, and the third is on the edge of the nearby forest. Other crosses are located around the village in different ways.

### 7.3 Discussion

In Central Europe, small sacral elements such as chapels, roadside crosses, shrines, and wayside chapels have deep historical and cultural roots [32]. These structures often date back to medieval times and were erected for various reasons, including religious devotion, to mark significant events, or as expressions of gratitude or protection. They are often situated at crossroads, village entrances, or along pilgrimage routes, serving as spiritual and physical landmarks [33].

These sacral elements play a crucial role in fostering a sense of community identity and continuity. They connect to the past, preserving local traditions and stories passed down through generations. Communities often take pride in maintaining and restoring these structures, which can become focal points for local festivals, processions, and gatherings, reinforcing social cohesion and collective memory [34].

For many people, these small sacral elements offer spaces for personal reflection, prayer, and contemplation, contributing to spiritual and psychological well-being. They provide comfort and protection as a reminder of the community's shared beliefs and values [35]. This spiritual dimension can be particularly significant in rural areas, where such elements often punctuate the landscape, integrating seamlessly into the residents' daily lives [36].

Sacral elements often contribute to the aesthetic and environmental quality of the landscape. They are frequently surrounded by trees, flowers, or other forms of natural beauty, enhancing the area's visual appeal. Additionally, these sites can become small-scale conservation areas, preserving local biodiversity and serving as habitats for various species [33, 37].

The presence of well-maintained sacral elements can also have a positive economic impact through tourism. Pilgrimage routes and historical trails that feature these structures attract visitors interested in cultural heritage, history, and religious tourism. This influx of visitors can benefit local communities economically through the hospitality industry, including lodging, dining and local crafts [38].

In modern times, these sacral elements face challenges such as neglect, environmental degradation, and the encroachment of urbanisation. However, they also present opportunities for community-driven conservation efforts [39]. Engaging residents, especially the younger generation, in preserving and promoting these sites can foster a deeper appreciation for cultural heritage and encourage sustainable development practices [40].

## 8. Conclusion

In both analysed territories, it was found that most of the monuments of small folk architecture are crosses lining the roads. This is a relict from the past when roads were lined with small crosses, which had religious functions, served as orientation points in the landscape, and as places of rest for travelling people due to the shadow of trees, which grew near the crosses. Except for the religious monuments, such as a cross, torments of God and small chapels, in the landscape are many memorials, especially to the victims of world wars. The least represented are monuments for orientation, such as signposts, milestones or border stones. These monuments were often removed when building reinforced paths were built. Many of the mapped monuments of small folk architecture are damaged because of the low financial funds required to repair historical monuments outside the central circle of interest.

Small sacral elements in the landscape of Central Europe play a multifaceted role in community development. They are religious symbols and integral components of cultural identity, social cohesion, environmental stewardship and economic activity. By recognising and preserving these structures, communities can maintain a vital link to their heritage while promoting sustainable development and well-being.

## Acknowledgements

This work could be elaborated thanks to support from the Faculty of Agriculture and Technology of the University of South Bohemia in Ceske Budejovice. Freely available materials and professional publications were used as sources.

## Thanks

Special thanks go to colleagues from the Faculty of South Bohemia of the University of South Bohemia in Ceske Budejovice and Ing. Marek Novotny, who co-worked on field surveys and subsequent data processing.

## Author details

Vendula Moravcova*, Jana Moravcova and Denisa Pekna
Faculty of Agriculture and Technology, University of South Bohemia in Ceske Budejovice, Czech Republic

*Address all correspondence to: moravv@fzt.jcu.cz

## References

[1] Claval P. Reading the rural landscapes. Landscape and Urban Planning. 2005;**70**(1-2):9-19

[2] Varela P. Archeology of a Medieval House. 1st. ed. Pilsen: West Bohemian University in Pilsen; 2004. 437 p; ISBN 80-903412-1-7

[3] Palang H, Printsmann A, Gyuro EK, Urbanc M, Skowronek E, Woloszyn W. The forgotten rural landscapes of central and Eastern Europe. Landscape Ecology. 2006;**21**:347-357

[4] Pesta J. Surface Survey of Folk Architecture and Rural Settlements. 1st. ed. Prague: National Heritage Institute; 2014. 174 p; ISBN 978-80-86516-78-3

[5] Lešková A, Vaishar A. Analysis of rural identity in terms of preservation of traditional folk architecture in the Vranov region. Public Recreation and Landscape Protection. 2020;**31**:6-15

[6] Syrovy B. Architecture - Testimony of the Times: Overview of Building Development and Architecture. Prague: State Publishing House of Technical Literature; 1974. 447 p.; ISBN 04-720-74

[7] Zelinsky W. The uniqueness of the American religious landscape. Geographical Review. 2001;**91**(3):565-585

[8] Stankova J, Sedlakova R, Posva R, Vodera S. Architecture in the Transformations of Millennia. 1st. ed. Prague: Sobotales; 2005. 303 p.; ISBN 80-86817-10-5

[9] Skabrada J. Folk Buildings: Architecture of Czech Countryside. 1st. ed. Prague: Argo; 2003. 239 p.; ISBN 80-7203-082-5

[10] Woods M. Rural geography: Processes, responses and experiences in rural restructuring. Rural Geography. 2004;**1**:1-352

[11] Dudak V, Neskudla B, Pošva R. World Encyclopedia Architecture: From Menhir to Deconstructivism. 1st. ed. Prague: Baset; 2000. pp. 535-1029. ISBN 80-86223-06-X

[12] Sadlo J, Pokorny P, Hajek P, Dreslerova D, Cilek V. Landscape and Revolution: Significant Breakthroughs in the Development of Cultural Landscape Czech Lands. 1st. ed. Prague: Mala Skala; 2005. 247 p. ISBN 80-86776-02-6

[13] Hajek T, Bukacova I. Story of Small Monuments: (from Disinterest to Fascination). 2nd. ed. Lomnice up. Popelkou: Studio JB; 2006. 140 p. ISBN 80-900903-9-7

[14] Christie N, editor. Landscapes of Change: Rural Evolutions in Late Antiquity and the Early Middle Ages. London: Routledge; 2017. 342 p. ISBN 9781315250915

[15] Liutikas D. Sacred space in geography: Religious buildings and monuments. In: Geography of World Pilgrimages: Social, Cultural and Territorial Perspectives. Cham: Springer International Publishing; 2023. pp. 77-112

[16] Klvač P, Musil L. About the memory of crosses, God's muc and chapels. In: Veronica - Magazine for Nature and Landscape Protection. Vol. XVII, num. 5. Brno: Czech Union for Nature Conservation; 2003. pp. 17-20. ISSN 1213-0699

[17] Kvet R. The Soul of the Landscape: Old Trails in the Changes of Ages. 1st. ed.

Prague: Academia; 2003. 195 p. ISBN 80-200-1012-2

[18] Stilgoe JR. Landscape and Images. Virginia: University of Virginia Press; 2005. 368 p. ISBN 978-0813937533

[19] Ismagilova G, Safiullin L, Gafurov I. Using historical heritage as a factor in tourism development. Procedia-Social and Behavioral Sciences. 2015;**188**:157-162

[20] Gaba Z, Hanzl Z, Prochazka L, Sedlicka K, Slouka J, Traxler J. Stone in Handmade Production of Czech Countryside. 1st. ed. Prague: Lidové noviny publishing house; 2003. 262 p. ISBN 80-7106-536-6

[21] Kyselka I. The importance of small landscape elements, experience with their recovery in the Czech Republic and abroad. In: Dejmalova K, editor. The Face of our Country: Home Landscape. 1st. ed. Lomnice nad Popelkou: Studio JB; 2001. 229 p. ISBN 80-900903-8-9

[22] Schulz H-D. Bildstöcke, Heiligenhäuschen und Kreuze – Christliche Wegebilder in und um Werne. Werne: Stadt Werne; 2003. p. 64

[23] Amster MH. Portable potency: Christianity, mobility and spiritual landscapes among the Kelabit. Anthropological Forum. 2009;**19**(3):307-322

[24] Turalija A, Vrsaljko A, Juric G, Kelava I, Drenjancevic M, Stanisavljevic A, et al. Small sacral objects as a landscape determination in Croatia. Bulletin of the University of Agricultural Sciences and Veterinary Medicine Cluj-Napoca Horticulture. Cluj-Napoca: University of Agricultural Sciences and Veterinary Medicine of Cluj-Napoca. 2011;**68**(1):437-443

[25] Tilke K, Bildstöcke H. Andachtgrotten. In: Hesener P, editor. Heggen im Wandel der Zeiten. Heggen: Schützenverein Heggen 1867; 1997. 680p

[26] Timmermann A. Highways to heaven (and hell): Wayside crosses and the making of late medieval landscape. The Authority of the Word. 2012;**1**:385-441

[27] Mecchi C. Kleinbauten im öffentlichen Raum III: Sakrale Kleinbauten. In: Merkblätter des Bundesamtes für Bevölkerungsschutz, Kulturgüterschutz. Bern: Bundesamt für Bevölkerungsschutz; 2007. 4 p

[28] Urfus V, Vit J, Wieser S, Karel T, Procházka Z. Stone Cross of Bohemia and Moravia. 2nd. ed. Prague: Argo; 2001. 475 p. ISBN 80-7203-370-0

[29] Vondruskova, A. and Vondruska, V. Church. 1st. ed. Prague: Vysehrad, 2014. 199 p. ISBN 978-80-7429-441-9

[30] Łach J, Marek A. The role of open-air calvaries in the context of landscape heritage protection of the eastern Sudetes. Geojournal of Tourism and Geosites. 2024;**53**(2):489-501

[31] Podovei PR. Monuments of folk architecture from Botoşani County. The Church of the Transfiguration. Romanian. Journal of Modern History. 2013;**4**(02):5-11

[32] Balikci G. Hemshin Folk Architecture in the Akbucak, Ortayol and Uërak Villages of the County of Pazar in Rize. London: Routledge; 2007. 345 p

[33] Cirtautas M, Alistratovaitė-Kurtinaitienė I, Dijokienė D. Lithuanian folk architecture research as a repository for the study of urban form. Architecture and Urban Planning. 2022;**18**(1):1-9

[34] Baniowska-Kopacz E. Folk architecture of Podhale. Ethnologia Polona. 2014;**35**:79-99

[35] Todorović M, Bjeljac Ž. Rural tourism in Serbia as a concept of development in undeveloped regions. Acta geographica Slovenica. 2009;**49**(2):453-473

[36] Nilson T, Thorell K. Cultural Heritage Preservation: The Past, the Present and the Future. Halmstad: Halmstad University Press; 2018. 158 p. ISBN 978-91-87045-94-3

[37] Schmitz S, Pepe L. Does local rural heritage still matter in a global urban world? In Sustaining Rural Systems. 2023;**1**:31-48

[38] Stubbs JH, Makaš EG. Architectural Conservation in Europe and the Americas. London: John Wiley & Sons; 2011. p. 512

[39] Dzikowska A, Zaręba A, Krzemińska A, Pawłowski K. The cultural landscape of rural cemeteries on the polish–Czech borderlands: Multifaceted visual analysis as an element of tourism potential assessment. Sustainability. 2023;**15**(18):13730

[40] Holland C. St Peter-on-the-Wall - the Last of Essex: Contemporary Architecture and Cultural Landscape. London: UCL Press; 2023. 332 p. ISBN 978-1-80008-437-7

Chapter 7

# The Role of Municipal Ecological Structure in Landscape Preservation

*Raquel Carvalho and Sérgio Lousada*


## Abstract

The Portuguese legal system includes in the municipal urban planning instruments an institute designed to link the use and occupation of space with environmental protection. Through the municipal ecological structure, municipal master plans contribute to the integration of green infrastructures into planning. The municipal ecological structure thus favors the maintenance of ecological balance and the protection, conservation and enhancement of the environment, landscape, and natural heritage in rural and urban areas. The content of municipal ecological structures differs according to geographical location, population density, and economic activities. Regardless of these natural variations, the main objective remains the same: to contribute to the achievement of SDGs, with specific emphasis on the goal of ensuring more sustainable and resilient cities considering the pernicious effects of climate change.


**Keywords:** green infrastructure, landscape, municipal ecological structure, SGD, sustainability, urban planning

## 1. Introduction

The intention of this chapter is to illustrate how law, particularly urban planning law, can contribute to landscape sustainability. One of the purposes of urban planning, and at the same time its premise, is the mapping of territorial circumstances: identifying geographical conditions, building areas (resulting in soil impermeabilization), economic activities, and demographic spread. Changes in land use, therefore, have an impact on the landscape, affecting environmental and socio-economic systems. Monitoring these changes is important for adjusting planning instruments, using a range of methodologies, and justifying the legal rules that require their periodic revision [1–3].

In Portugal, land management instruments incorporate the protection of the landscape and include a tool specifically devoted to this purpose: the ecological structure. It can be found in the various planning instruments, but due to the general binding nature of municipal plans (self-binding and hetero-binding), we will focus on Municipal Master Plans.

What is analogous to this planning instrument in all municipal plans is the environmental protection dimension. In addition to having to incorporate the soil's situational dimensions from other instruments, the plans incorporate different green (and blue) infrastructures into the municipal ecological structure (MES hereinafter). Each municipality has its own features [1], thus its own urban planning solutions. Portugal has many municipalities, despite its small size. Although the figure of districts has organically disappeared, it is still useful as a criterion for territorial distribution, for instance, electoral districts. Thus, there will be examples of municipalities from north to south, from inland to coastal, which will enable the analysis of the most important MES. Still, even though district capitals are usually the larger municipalities, demographic distribution and territorial dimension are quite dissimilar, which may affect the full reflection of the use of the instrument. Portugal also has two island autonomous regions, whose geographical location means that urban planning must take account of the circumstance of insularity. As Lousada et al. state "island territories suffer unique challenges and are particularly vulnerable to change due to their geographic situation" [4].

Nature-based Solutions (NBS) is an elusive, extensive non-legal concept that encompasses solutions inspired by nature. It can embed nature solutions resorting to water, vegetation, or both. NBS is a set of solutions to strengthen urban resilience, by minimizing, or solving, some of the challenges posed by climate change.

There are several types of solutions and combinations. Kimic & Ostrysz [5] describe 19 Blue Green Infrastructures (BGI) solutions considering their location on the runoff troughs, grassed swales, infiltration trenches, vegetated swales, (street-side) bioretention basins, grassed retention and infiltration basins, rain gardens, wetland ponds", surface water reservoirs, retention and infiltration water reservoirs, water squares, permeable/pervious pavements, undergrounded (infiltration wells, infiltration boxes, structural tree root cells, underground water reservoirs) or above surface (blue roofs, green roofs, green walls) [5, 6].

Green Infrastructures (GI) are one of the tools used in MES. Due to the difference between the municipalities, it is relevant to identify the most used GI in the Municipal Plans of district capitals. However, when presenting the GIs, they will be accompanied by examples from other locations to reinforce their usefulness and the effects already achieved, particularly in the pursuit of the 11th and 13th SDG [7].

## 2. The Portuguese legal framework

### 2.1 Introduction

The literature acknowledges the key role of local governments in tackling adverse climate change effects through urban law, both in its planning and management dimensions [8–14], and in pursuing sustainability through urban planning within European hard and soft Law [15, 16] and the connection to the implementation of the UN SDG [2, 5, 17, 18]. It should be stressed, though, that top-down imposed solutions are not adequate, given the heterogeneity of the territories and cities, their size, geographical location, and climate circumstances [19]. The solutions must be tailored-made [20, 21]. Additionally, citizens' – as well as other stakeholders' – involvement through collaboration procedures is an important principle for both public urban policy and urban law [12, 22–27], reinforcing social actors' support for the inclusion of NBS in urban policies [28]. Still, mere local approaches without State support may not be enough [8]. Finally, NBS monitoring and assessment included in urban planning contribute to social awareness of the respective benefits, fostering the mobilization of society to support the inclusion of these solutions [29–35].

### 2.2 The Portuguese local government and urban planning

In Portugal, besides the Constitution, which relates sustainability to urban planning and building (article 66 no 2), there are some relevant macro legal solutions. The Environment Framework Law – EFL (Law no 19/2014, 14th April), the Climate Framework Law (Law no 98/2021, 31st December), and the Waste Law (Decree-Law no 102-D/2020, 10th December), all enshrine sustainable solutions with reflections on urban planning law. Then, at a dedicated level, the Public Policy Framework Law on Land Use, Land Planning and Urban Planning – PPFLLUP (Law no. 31/2014, 30th May) and the legal framework for territorial management instruments (Decree-Law no 80/2015, 14th May) particularly detail that connection. There is also an important regulation regarding building requirements (Decree-Law no 555/99, 16th December).

Urban planning is a legal instrument that defines the use, the occupation, and the transformation of land, whether urban or rural. It is therefore an insurmountable tool when the objective is to make the human habitat more sustainable and cities resilient (11th SDG) [36–38].

The EFL establishes the interdisciplinary and transversal character of the planning instruments (article 13). The Climate Framework Law, in turn, relates green taxes to urban sustainability (article 28), considers local governments as the promoting entities of sustainable management of urban spaces (article 51), and foresees State intervention with forest and green spaces to mitigate the heat island effect and promote green roofs and walls (article 57). In fact, this planning tool crosscuts all the main public interests: defense, public health, foreign policies, food, water resources, forest, energy, urban mobility... The principles enshrined in the Portuguese Climate Framework Law are very similar to those described by Monteiro, Ferreira, & Antunes, [24]: sustainable development, citizens' participation in planning instruments, articulation, integration, prevention, precaution, accountability (article 4).

The PPFLLUP also underlines sustainable development as a relevant instrument for climate action. The Portuguese legal framework is aligned with the European Green Deal and The Action Plan for the Circular Economy. It establishes the articulation principle among planning instruments (article 45) [39, 40] the planning revision principle grounded in "evolution or reconsideration of the economic, social, cultural and environmental conditions underlying their elaboration" (article 50). Furthermore, it embodies the green spaces in urban planning goals (article 2), as citizens' right and duty to preserve (articles 6 and 7); and the administrative bodies' duty to preserve green spaces as an active power (article 8). In Portugal, territorial management instruments are various and distributed, in terms of responsibility for their preparation, among the various levels of Public Administration. Portuguese law classifies such instruments in the light of various criteria, the most relevant being that of external binding nature, in the context of spatial planning and urban planning. The legal framework for territorial management instruments (Decree-Law no 80/2015, 14th May) details the preparation, amendment, and review of planning instruments. One of the most important planning instruments that binds citizens is the Municipal Master Plan (MPP), whose elaboration is mandatory. It "defines the strategic framework for territorial development of the municipality" (article 27). Therefore, local authorities have a fundamental role to play in the context of the use of NBS, which has room in the city space as a community but also in citizens' buildings. These requirements (the distribution of services and economic activities in the city space and the requirements for the construction of buildings) should be embodied in MMP.

In our perspective, this instrument is particularly adequate for collecting all the requirements that the literature has been referring to as being fundamental.

First, because of the principles associated with this legal instrument. The Portuguese law establishes the principle of citizen participation in the elaboration of MMP through public consultation, thus summoning citizens to collaborate in the definition of solutions for their city space and on the more specific requirements of the buildings they inhabit (articles 6 and 89) [40]. In other words, the intention is that everyone contributes on a macro (city space) and micro (their dwelling) level to mitigate the adverse effects of climate change with these solutions. When consulting the material content of MMP (article 96), one can find the cross-cutting feature among urban planning and economy, transportation, culture, health, safety, energy, natural resources, drainage, waste, sustainability, urban rehabilitation... The requirements for building also contemplate some rules regarding the duty to give away space to establish green urban spaces (article 43, for instance). But the specific requirements regarding buildings are set in local government regulations.

Second, as in Portugal the MMP has a regulatory nature, the law currently requires a cost-benefit study to be carried out for the preparation of any regulation. Therefore, the incorporation of the analysis proposed by the literature regarding the inclusion of the NBS will not constitute, per se, a novelty to be incorporated by the legislator.

The Portuguese legislation can be improved through the incorporation of NBS implementation monitoring and evaluation mechanisms [41] (and even others that would also benefit therefrom), fulfilling the provisions of the CFL. It should be noted that, given the evolution of cities and demographic movements, Portuguese law already requires the cyclical re-examination of MMP solutions through review mechanisms (articles 115 and 124). On the other hand, considering the existence of other planning instruments, it is necessary to adapt MMPs and bring them into line with hierarchically superior planning instruments.

Finally, due to the articulation principle, there is room to expressly recognize multifunctionality as an urban planning principle [25, 27, 42]. Hansen et al. [41] stress that the "principle of multifunctionality alone will not be able to secure sufficient provision of green spaces in densifying cities but needs to be combined with other principles such as the conservation of valuable green spaces, securing green space quantity and quality, and increasing connectivity and the accessibility of green infrastructure" [43, 44].

### 2.3 The municipality ecological structure (MES)

The Portuguese urban law planning foresees MES as an important tool to address the need to protect environment issues. Article 75 of Decree-Law no 80/2015, 14th May, foresees MES as a part of MMP. Then, Article 99 also foresees it regarding Urbanization Plans. Article 107 as well, related to the Detail Plan. The Urbanization Plan develops and materializes MP and the Detail Plan "develops and concretizes in detail the proposals for the occupation of any area of municipal territory".

The Government has issued a Decree (regulation, not with legal value), where the MES is described (Regulatory Decree no. 5/2019, 27th September). In the Annex of the regulation's operative part, several urban concepts are established, and MES is one of them. File 1–29 states: "The municipal ecological structure is a set of land areas which, by virtue of their biophysical, cultural or landscape characteristics, their ecological continuity and their organization, have the main function of contributing to ecological balance and to the protection, conservation and environmental and landscape improvement of rural and urban areas".

One of the main features is continuity between rural and urban areas. Nevertheless, in each type or area, the role and ecological design may vary:

a. rural area: "the municipal ecological structure involves the areas of land assigned to the fundamental nature conservation network in the municipality's territory, natural areas subject to risks and vulnerabilities and other areas of land that are selected and delimited according to the municipal interest, namely for environmental, landscape and natural heritage protection and enhancement reasons and ecosystem services".

b. urban areas: "[the municipal ecological structure] includes green spaces for collective use and other spaces, of a public or private nature, which are necessary for the balance, protection and enhancement of the environment, landscape and natural heritage of the urban area, particularly with regard to: (i) Regulation of the hydrological cycle (preservation of soil permeability and creation of retention areas, within the framework of urban flood prevention); (ii) Bioclimatic regulation of the city (reducing thermal amplitudes and maintaining air humidity levels); (iii) Improving air quality (reducing the concentration of atmospheric pollution in urban centers); (iv) Biodiversity conservation (maintaining habitats); (v) Ecosystem services". According to the definition enshrined in Article 43 of Decree-Law 555/99, December 16th, within MES, green spaces for collective use include parks and gardens.

## 3. Nature-based solutions (NBS) and their role in landscape

There are many descriptions of NBS:

a. United Nations (UN hereinafter): NBS "are actions to protect, conserve, restore, sustainably use and manage natural or modified terrestrial, freshwater, coastal and marine ecosystems which address social, economic and environmental challenges effectively and adaptively, while simultaneously providing human well-being, ecosystem services, resilience and biodiversity benefits, and recognizes that nature-based solutions". Thus aligning NBS with SDG and stating that "may contribute significantly to climate action, while recognizing the need for analysis of their effects, including in the long term, and acknowledging that they do not replace the need for rapid, deep and sustained reductions in greenhouse gas emissions, but can improve action for adaptation and resilience to and mitigation of climate change and its impact" [45].

b. European Commission: NBS "are inspired and supported by nature, which are cost-effective, simultaneously provide environmental, social and economic benefits and help build resilience. Such solutions bring more, and more diverse, nature and natural features and processes into cities, landscapes and seascapes, through locally adapted, resource-efficient and systemic interventions" (https://

research-and-innovation.ec.europa.eu/research-area/environment/nature-based-solutions_en accessed 10th June 2024) (**Figure 1**).

The European Union has several financed programs regarding NBS linked to urban and landscape planning. It is useful to illustrate some of these solutions already in place.

a. Within the "Re-naturing urbanization", that aims to "cover vast urban areas and mitigate the effects of climate change", the following NBS can be found: green routes, shade trees, colling trees, arboreal areas, parklets, urban carbon sink, green resting areas.

b. there are also "singular green infrastructures", such as Hydroponic green façades, green roofs and walls, Urban Garden Bio-Filters, Green filter areas, green shady structures, Electro wetlands, and vertical GI (vertical mobile gardens, green noise barriers, green fences).

c. regarding water resources, there is also a wide set of NBS: channel re-naturalization, cool pavements, cycle-pedestrian green paths, green pavements, hard drainage pavements, natural wastewater management, green filter areas, floodable parks, hard drainage-flood prevention, grassed swales and water retention pounds, urban catchment forestry, SUDs, and rain gardens.

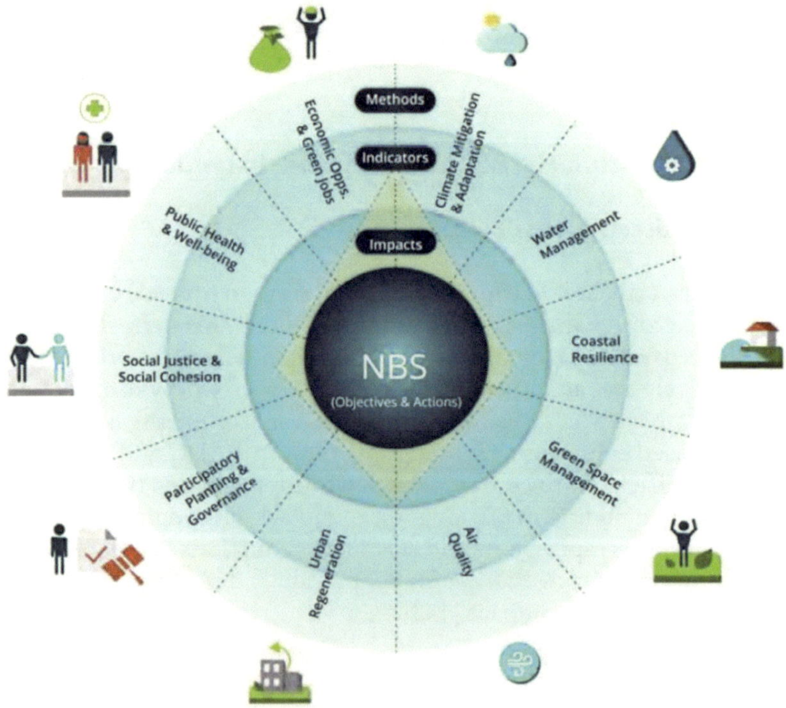


**Figure 1.**
*EKLIPSE challenges (D1.1 - NBS CATALOG - https://www.urbangreenup.eu/insights/deliverables/d1-1---nbs-catalogue.kl accessed 10th June 2024).*

Besides the role of these NBS, replacing the perception of the gray, built-up humanized landscape with the smooth, greener landscape, NBS also have a preeminent role in softening some of the most harmful climate change effects (13th SDG), such as the heat island effect, preventing droughts and floods [46–52], contributing to temperature regulation in public spaces [5, 41, 47, 53–58]. Thus, improving citizens' (individual and collective) health [59] and air quality [60–62].

The European Commission recognizes GI as "a successfully tested tool for providing ecological, economic and social benefits through natural solutions", which provides cheaper and more durable solutions in spatial planning and territorial development than constructed ones. Assuming their complexity and diversity (as NBS have been described as an umbrella concept [11, 25, 27, 63–67]), GI is defined as a "strategically planned network of natural and semi-natural areas with other environmental features designed and managed to deliver a wide range of ecosystem services. It incorporates green spaces (or blue if aquatic ecosystems are concerned) and other physical features in terrestrial (including coastal) and marine areas. On land, GI is present in rural and urban settings". The positive outcomes of using GI cover not only ecological and environmental aspects but also health-related issues, linked to air pollution, the spread of vector-borne diseases, and community ties and networking, related to public spaces' leisure time. Furthermore, it can create "opportunities to connect urban and rural areas and provides appealing places to live and work in", associating the issue of food production through community gardens. They are recognized to be effective in fighting and/or mitigating climate change effects [68]. GI are also foreseen in "EU Biodiversity Strategy for 2030", clearly related to urban planning and management [69]: "This strategy aims to reverse these trends and stop the loss of green urban ecosystems. The promotion of healthy ecosystems, green infrastructure and nature-based solutions should be systematically integrated into urban planning, including in public spaces, infrastructure, and the design of buildings and their surroundings" [70].

The introduction of GI is related to and has different impacts depending on the territory. It may involve more land use changes in continental territories, which are not delimited by a maritime border, than in island territories. The processes of urbanization and cities' growth are much more constrained in the latter than in the former, and the influence on the landscape is therefore equally different [4].

## 4. Portuguese municipalities: the case study

As already stated before, Portugal is a small country but has 308 municipalities. All of them have MMP approved, even though some of them have already been revised once or even twice (**Figure 2**).

Referring to urbanization and detailed plans not only of them have them. According to official sources (General Direction for Territory https://reot.dgterritorio.gov.pt/sistema-de-gestao-territorial/pu-e-pp accessed 10th June 2024) (**Figure 3**).

### 4.1 The municipality ecological structure (MES) in Portugal district's capital

There are 18 district capitals in continental Portugal.

From the analysis of all 18 MMP, only Viana do Castelo (North and Coastal Municipality) and Guarda (Interior Municipality) have no information regarding MES. The first is due to the revision of MMP and Preventive Measures in force; the latter does

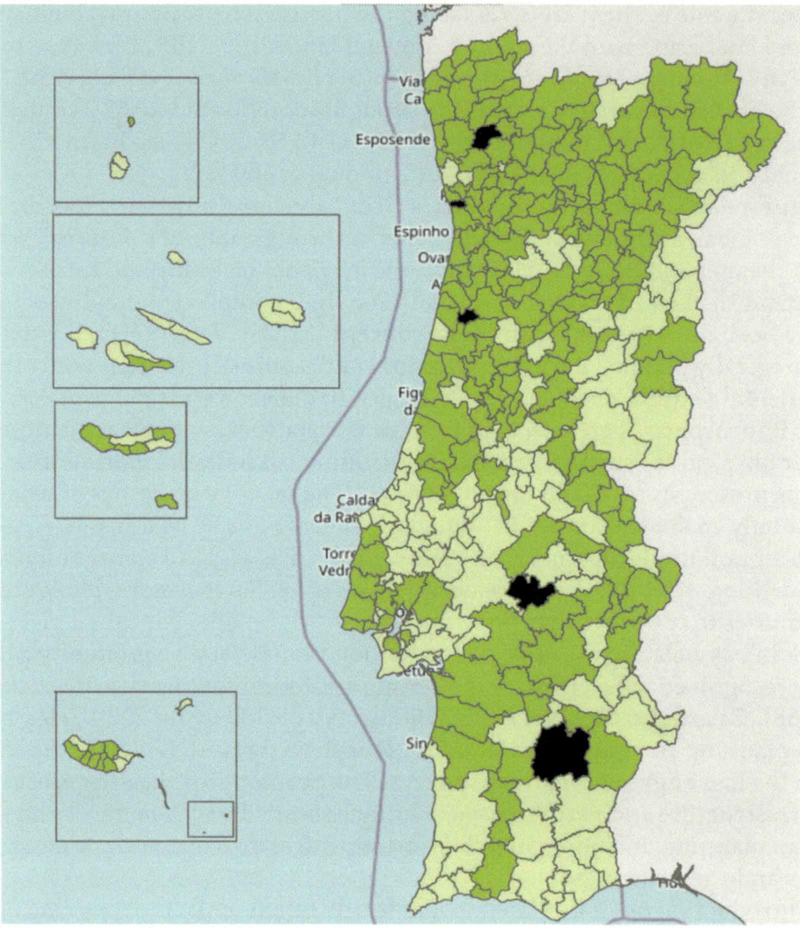


**Figure 2.**
*Municipalities in Portugal – MMP in force by 2022 (https://reot.dgterritorio.gov.pt/sistema-de-gestao-territorial/pdm). Light green (1st publication – 114); dark green (1st revision – 189); black (2nd Revision – 5).*

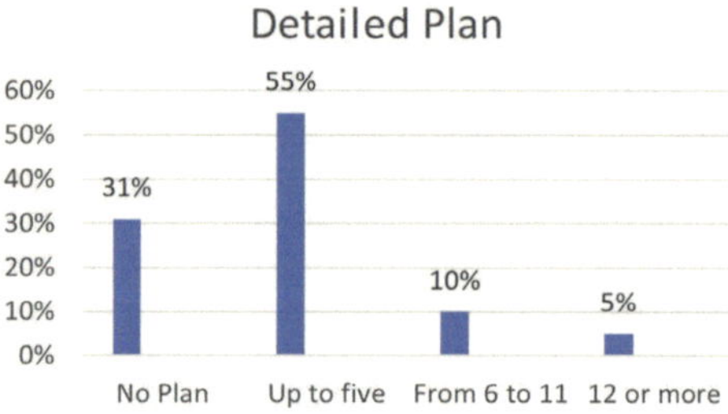


**Figure 3.**
*Number of detailed plans.*

not foresee the tool. There are two Municipalities – Castelo Branco e Santarém – that do not mention MES, but have rules regarding "protection areas" and "green spaces".

Almost all MMP enshrine the continuity principle (Aveiro, Coimbra, Évora, Faro, Leiria, Lisboa, Portalegre, Porto, Setúbal, Vila Real e Viseu). Thus, Municipalities from inland and coastland, north and south.

*DOI: http://dx.doi.org/10.5772/intechopen.1006045*

Not all MES are designed in the same way. This urban planning tool, as urban planning in general, must be tailor-made [1]. Therefore, even though the elements and goals are similar, the concrete design varies. The constituents are, in general, areas identified as "the most important biophysical values", the "fundamental ecosystems based on the consideration of different biotopes", areas that aim to contribute to the ecological balance and to the protection, conservation and environmental and landscape enhancement of the territory, cultural values and resources, with continuity, considering urban and rural land, and "the protection, conservation and environmental and landscape enhancement".

Considering the tailored-made design:

a. Aveiro: "Natural Spaces, Green Spaces, Agricultural Spaces and Forest Spaces, and it also incorporates the territorial reference of the homogeneous subregions and the ecological corridor defined in the Centro Litoral Regional Forest Management Program (PROF CL)"; (this municipality is coastland located and has both urban and rural land);

b. Beja: "It is divided into two areas, depending on the nature of the space, namely: Urban Soil MES and Rural Soil MES"; (this municipality is inland located and mainly rural land);

c. Braga: "Fundamental Ecological Structure (FES) and the Urban Ecological Structure (UES)"; (this municipality, even though it is not near the coast, is mainly urban);

d. Bragança: "Agricultural Areas, Type I Agro-sylvo-pastoral Areas, Conservation and Protection Forest Areas, Natural Areas and Land allocated to the Urban Ecological Structure, as well as the remaining areas integrated into the National Ecological Reserve that are not covered by classified land"; (this municipality is inland located and rural);

e. Évora: "a) Core areas - Consisting of the areas classified as Sites of Community Interest - Monfurado and Évora Special Protection Area; b) Ecological connectivity areas/ecological corridors - Consisting of the main watercourses and their riparian galleries; the main headwaters of watercourses; areas threatened by floods; aquifer recharge; the municipality's reservoirs and their respective buffer strips; areas with high or very high erosion risks and/or already degraded by intense erosion processes; areas corresponding to aquifers that are highly susceptible to pollution; patches of holm oak, cork oak and mixed forests; thermophilic scrub and woodland and riparian ecosystems";

f. Faro: "Natural and Landscape Spaces, Green Spaces and Agricultural Spaces, Structuring and Connectivity Corridors and also the core areas of the Regional Environmental Protection and Enhancement Structure (ERPVA) defined in the PROT Algarve";

g. Leiria: "a) Fundamental areas - corresponding to areas of the territory with high natural value that are strategic in terms of preservation, sustainability and ecological continuity; b) Complementary areas - these cover areas of the territory with natural values and/or whose biophysical characteristics play an important role in the ecological balance and in the protection, conservation and environmental and landscape enhancement of rural and urban land and relate to urban green spaces; c) Ecological corridors - integrate the structuring and complementary corridors and correspond to areas of the territory whose main function is to ensure connectivity between the main ecological systems";

h. Portalegre: "Rustic soil with two components: Fundamental municipal ecological structure; Complementary municipal ecological structure; the urban soils included in the spaces allocated to the urban ecological structure"; (this municipality is inland located and rural);

i. Setúbal: "Fundamental Ecological Structure, which is made up of the most important areas for the natural systems to work and integrates the areas that support the fundamental ecological systems and whose protection is essential for the sustainable functioning of the territory; Urban Ecological Structure, which aims to enhance and intensify ecological processes in built-up areas, thus constituting a structure for protection, climate regulation and support for plant production integrated into the urban fabric."; (this municipality is coastland located and urban);

j. Vila Real: "a) Natural environment areas; b) Green and community use areas; c) Green areas for protection and safeguarding; d) Green framework areas; e) Mixed green areas".

The main concerns which can be found in MMP and the specific tools (NBS, mainly GI) are:

a. protection of native flora;

b. fighting the heat island effect by afforesting streets and protecting urban parks and gardens;

c. concern about river flooding and protecting riverbanks with green solutions;

d. small surfaces to be impermeable;

e. "The design of new gardens and urban parks should promote their resilience, using permeable paving, terrain modelling that allows infiltration in situ and a vegetation structure adapted to local soil and climate conditions, with a focus on reducing installation and maintenance costs and contributing to increased biodiversity" (Coimbra);

f. "Promote the ecological continuity and articulation of the different spaces, based on existing or to be created natural physical elements, namely water lines, green spaces and tree-lined streets" (Faro);

g. "ensuring the continuity and complementarity of natural systems in the urban area, the ecological and physical sustainability of the environment, the functions of biological systems, biodiversity, the control of water flow and wind circulation, bioclimatic comfort and the enhancement of the landscape heritage" (Lisboa);

h. "The wet system includes the areas relating to open drainage lines, adjacent areas, rainwater retention basins, water resurgence areas, alluvial areas and areas subject to flooding" (Lisboa);

i. "Atlantic and riverfront areas, watercourses and retention basins and alluvial areas that make up the wet system, as well as sloping areas; priority should be given to soil permeability, safeguarding the hydrological cycle, protecting and enhancing natural resources, restoring the riverside ecosystem, promoting the riparian gallery and implementing sustainable drainage systems; (...) the use of endemic flora, the maintenance of ecologically valuable plant stands, the adoption of optimised water and energy efficiency solutions, using natural-based solutions whenever possible" (Porto);

j. "climate regulation (in order to limit extreme phenomena of temperature, cold and heat, allowing a more favourable climate for the population, namely through green spaces, tree-lined roads and bodies of water); air regulation (the extraction of gaseous substances from the atmosphere that are harmful to human health, derived from emissions from transport, industry and housing, providing the dissemination of good quality air through natural ecosystems); water regulation (extraction of substances from water bodies, derived from discharges of pollutants, ensuring and increasing the quality of water with an impact on the recharge of aquifers, the maintenance of water bodies and the supply of drinking water); control of extreme events (acting through ecosystems to mitigate the harmful effects of abnormal natural events, in particular floods, coastal erosion, heat island phenomena, forest fires and droughts); landscape value (taking landscape into account as a factor in human well-being and the quality of life of the population in general). (...) Encourage the harvesting and storage of rainwater and its reuse in irrigation systems for public or private green spaces; use of native plant species; creation of public spaces for urban horticulture; encourage the planting of vegetation adapted to the Mediterranean climate; mitigate the effect of urban heat islands, namely by planting tree and shrub vegetation in streets and public spaces; Using plant species with a higher carbon capture capacity; Creating retention basins upstream of urban settlements; Freeing up areas around water lines, flood beds and areas at risk of tsunamis and floods; Encouraging an increase in permeable areas on urban land and restricting waterproofing; Ensuring that rainwater is collected and correctly channelled." (Setubal).

k. usually, in all MES, construction is forbidden as the destruction of natural environment.

So, in Portugal's district capitals there are already several NBS implemented, mainly GI such as afforesting streets and protecting urban parks and gardens; permeable paving, tree-lined streets, rainwater retention basins, water resurgence areas, alluvial areas, promoting the riparian gallery and implementing sustainable drainage systems, use of endemic flora, the maintenance of ecologically valuable plant stands, encouraging the harvesting and storage of rainwater and its reuse in irrigation systems for public or private green spaces, creation of public spaces for urban horticulture; creating retention basins upstream of urban settlements (**Figure 4**).

GI, besides the immediate goal, can also, as the continuity principle suggests, connect natural and artificial spaces, mainly in high-density built areas. This connection makes the landscape continuous and without sectioning [71].

**Figure 4.**
*Word cloud with the most common NBS concepts in the various plans.*

## 5. Conclusions

Urban planning has a relevant role in landscape preservation. The Portuguese case of municipal planning, within the bigger municipalities that meet the district capital criterion, has shown that nature is already used in urban planning. For that purpose, Portuguese urban law has a tool – the ecological infrastructure – that must be enshrined in many planning instruments, including Municipal Master Plans.

Even though not all of them have the same design, which is comprehensible due to the different territory they apply to, they share similar concerns and solutions: afforesting streets and protecting urban parks and gardens; permeable paving, tree-lined streets, rainwater retention basins, water resurgence areas, alluvial areas, implementing sustainable drainage systems, use of endemic flora, the maintenance of ecologically valuable plant stands, and the creation of public spaces for urban horticulture.

Studies have been carried out on land use change in island and continental territories over a long period of time. In island territories, such as Madeira, there has been "stagnation in the growth of its urban fabric and change in the main vegetation" [4]. The stagnation of urban growth cannot, however, be a constant in all island territories, because not all insularity is of the same type. Therefore, as in the Canary Islands, there can be urban growth that is even unrestrained [1]. In continental territories such as Galicia, human occupation of the land has meant a decrease in forest area [3]. Similar trends have been observed in Alto Tâmega, although without the strict dimension of forest [2].

## Conflict of interest

The authors declare no conflict of interest.

## Notes/thanks/other declarations

The corresponding author wishes to express her gratitude toward her coauthor and the editorial board for making this chapter financially possible.

## Appendix. Municipal master plans

Aveiro - https://www.cm-aveiro.pt/cmaveiro/uploads/writer_file/document/4117/regulamento_pdm_1revisao.pdf

Beja - https://geoportal.cm-beja.pt/documentos/IGT/PDM_Aviso_22168_2023.pdf

Braga - https://www.cm-braga.pt/archive/doc/RegPDMatualizado04082021.pdf

Bragança - https://www.cm-braganca.pt/cmbraganca2020/uploads/writer_file/document/64/20100715220602250332.pdf

Castelo Branco - https://www.cm-castelobranco.pt/media/3104/8%C2%AA-altera%C3%A7%C3%A3o_2017_republica%C3%A7%C3%A3o-integral-do-regulamento.pdf

Coimbra - - https://www.cm-coimbra.pt/wp-content/uploads/2023/06/Regulamento_AvisoDR_-3731_2022.pdf

Évora - https://www.cm-evora.pt/wp-content/uploads/2021/11/7.-AVISO-2174_2013.pdf

Faro - http://mapas.cm-faro.pt/geoportal/docs/pdm_2023/REGULAMENTO_Junho_2023.pdf

Guarda - https://www.ccdrc.pt/wp-content/uploads/2021/07/04_PDM_Guarda_3AltAdaptPOPNSE-public_30Jul2021-a37.pdf

Leiria - https://www.cm-leiria.pt/cmleiria/uploads/document/file/10034/1_correcaoMaterial.pdf

Lisboa - https://diariodarepublica.pt/dr/detalhe/declaracao-retificacao/703-2020-145585233

Portalegre - https://www.cm-portalegre.pt/wp-content/uploads/2022/09/aviso_dr_regulamento.pdf

Porto - https://portaldomunicipe.cm-porto.pt/documents/20122/437751/Regulamento_altera%C3%A7%C3%A3o+por+adapta%C3%A7%C3%A3o.pdf/ceef7c02-8625-2d07-1674-afc0c4c4aa4f?t=1642676889761

Santarém - https://www.cm-santarem.pt/images/santarem/servicos_municipais/ordenamento/1416_Regulamento%20PDM_Vigor.pdf

Setúbal - https://www.mun-setubal.pt/wp-content/uploads/2021/10/RPDMS_C_Regulamento.pdf

Viana do Castelo - https://www.cm-viana-castelo.pt/wp-content/uploads/2023/03/Aviso_5770_2023_Suspensao_parcial_PDM_Estabelecimento_medidas_preventivas.pdf

Vila Real - https://www.cm-vilareal.pt/images/areas_servicos/planeamento/PDM_marco_2018.pdf

Viseu - https://www.cm-viseu.pt/fotos/documentos_categorias_ficheiros/aviso_5793_2023_6299944176414ab2ab8cad.pdf

## Author details

Raquel Carvalho[1*] and Sérgio Lousada[2,3,4,5,6]

1 Faculty of Law, Universidade Católica Portuguesa, Porto, Portugal

2 Faculty of Exact Sciences and Engineering (FCEE), Department of Civil Engineering and Geology (DECG), University of Madeira (UMa), Funchal, Portugal

3 VALORIZA - Research Centre for Endogenous Resource Valorization, Portalegre, Portugal

4 RISCO - Civil Engineering Department of University of Aveiro, Aveiro, Portugal

5 CITUR - Madeira - Centre for Tourism Research, Development and Innovation, Madeira, Portugal

6 Research Group on Environment and Spatial Planning (MAOT), University of Extremadura, Badajoz, Spain

*Address all correspondence to: rmcarvalho@ucp.pt

## References

[1] Naranjo, Gómez JM, Lousada S, Garrido Velarde J, Castanho RA, Loures L. Land-use changes in the canary archipelago using the CORINE data: A retrospective analysis. Land. 2020;**9**. DOI: 10.3390/land9070232

[2] Lousada S, Naranjo Gómez JM, Loures L. The Evolution of Land-Use Changes in the Alto Tâmega Region, Portugal: From 1990 to 2018 - a Vision of Sustainable Planning. In: Sustainable Regional Planning. England: London, UKEd. IntechOpen; 2023. DOI: 10.5772/intechopen.110036

[3] Lousada S, Naranjo Gómez JM. Analyzing the evolution of land-use changes related to vegetation, in the Galicia region, Spain: From 1990 to 2018. In: Vegetation Dynamics, Changing Ecosystems and Human Responsibility. England: London, UKEd. IntechOpen; 2022. DOI: 10.5772/intechopen.106015

[4] Lousada S, Cabezas J, Castanho RA, Naranjo Gómez JM. Land-use changes in insular urban territories: A retrospective analysis from 1990 to 2018. The case of Madeira Island—Ribeira Brava. Sustainability. 2022;**14**. DOI: 10.3390/su142416839

[5] Kimic K, Ostrysz K. Assessment of blue and green infrastructure solutions in shaping urban public spaces-spatial and functional, environmental, and social aspects. Sustainability. 2021;**13**(19). DOI: 10.3390/su131911041

[6] Hanson HI, Wickenberg B, Alkan OJ. Working on the boundaries—How do science use and interpret the nature-based solution concept? Land Use Policy. 2020;**90**:104302. DOI: 10.1016/j.landusepol.2019.104302

[7] Schmidt S, Guerrero P, Albert C. Advancing sustainable development goals with localised nature-based solutions: Opportunity spaces in the Lahn river landscape, Germany. Journal of Environmental Management. 2022;**309**. DOI: 10.1016/j.jenvman.2022.114696

[8] Gantar D, Kozamernik J, Erjavec IS, Koblar S. From intention to implementation of vertical green: The case of Ljubljana. Sustainability. 2022;**14**(6). DOI: 10.3390/su14063198

[9] Cheng Z, Nitoslawski S, van den Bosch CK, Sheppard S, Nesbitt L, Girling C. Alignment of municipal climate change and urban forestry policies: A Canadian perspective. Environmental Science & Policy. 2021;**122**:14-24. DOI: 10.1016/j.envsci.2021.04.005

[10] Tzoulas K, Galan J, Venn S, Dennis M, Pedroli B, Mishra H, et al. A conceptual model of the social-ecological system of nature-based solutions in urban environments. Ambio. 2021;**50**(2, SI):335-345. DOI: 10.1007/s13280-020-01380-2

[11] Bush J, Coffey B, Fastenrath S. Governing urban greening at a metropolitan scale: An analysis of the living Melbourne strategy. Australian Planner. 2020;**56**(2, SI):95-102. DOI: 10.1080/07293682.2020.1739093

[12] Bush J, Doyon A. Building urban resilience with nature-based solutions: How can urban planning contribute? Cities. 2019;**95**. DOI: 10.1016/j.cities.2019.102483

[13] Cilliers EJ. Refleting on green infrastructure and spatial planning in Africa: The complexities, perceptions,

and way forward. Sustainability. 2019;**11**(2). DOI: 10.3390/su11020455

[14] Matthews T, Lo AY, Byrne JA. Reconceptualizing green infrastructure for climate change adaptation: Barriers to adoption and drivers for uptake by spatial planners. Landscape and Urban Planning. 2015;**138**:155-163. DOI: 10.1016/j.landurbplan.2015.02.010

[15] de Luca C, Naumann S, Davis M, Tondelli S. Nature-based solutions and sustainable urban planning in the European environmental policy framework: Analysis of the state of the art and recommendations for future development. Sustainability. 2021;**13**(9). DOI: 10.3390/su13095021

[16] Castellanos LA, Versini PA, Bonin O, Tchiguirinskaia I. A text-mining approach to compare impacts and benefits of nature-based solutions in Europe. Sustainability. 2020;**12**(18). DOI: 10.3390/su12187799

[17] Nita MR, Badiu DL, Onose DA, Gavrilidis AA, Gradinaru SR, Nastase II, et al. Using local knowledge and sustainable transport to promote a greener city the case of Bucharest, Romania. Environmental Research. 2018;**160**:331-338. DOI: 10.1016/j.envres.2017.10.007

[18] Bellamy CC, van der Jagt APN, Barbour Shelleyand Smith M, Moseley D. A spatial framework for targeting urban planning for pollinators and people with local stakeholders: A route to healthy, blossoming communities? Environmental Research. 2017;**158**:255-268. DOI: 10.1016/j.envres.2017.06.023

[19] Rey Mellado R, del Pozo SC, Franchini Alonso MT. Nature based solutions: Urban strategies for adaptation to climate change. Habitat Y Sociedad. 2021;**14**:243-262. DOI: 10.12795/HabitatySociedad.2021.i14.13

[20] Dorst H, van der Jagt A, Runhaar H, Raven R. Structural conditions for the wider uptake of urban nature-based solutions-a conceptual framework. Cities. 2021;**116**. DOI: 10.1016/j.cities.2021.103283

[21] Colleony A, Shwartz A. Beyond assuming Co-benefits in nature-based solutions: A human-Centered approach to optimize social and ecological outcomes for advancing sustainable urban planning. Sustainability. 2019;**11**(18). DOI: 10.3390/su11184924

[22] Arlati A, Roedl A, Kanjaria-Christian S, Knieling J. Stakeholder participation in the planning and Design of Nature-Based Solutions. Insights from CLEVER cities project in Hamburg. Sustainability. 2021;**13**(5). DOI: 10.3390/su13052572

[23] Malekpour S, Tawfik S, Chesterfield C. Designing collaborative governance for nature-based solutions. Urban Forestry & Urban Greening. 2021;**62**. DOI: 10.1016/j.ufug.2021.127177

[24] Monteiro R, Ferreira J, Antunes P. Green infrastructure planning principles: An integrated literature review. Land. 2020;**9**(12):525. DOI: 10.3390/land9120525

[25] Dorst H, van der Jagt A, Raven R, Runhaar H. Urban greening through nature-based solutions – Key characteristics of an emerging concept. Sustainable Cities and Society. 2019;**49**:101620. DOI: 10.1016/j.scs.2019.101620

[26] Frantzeskaki N. Seven lessons for planning nature-based solutions in cities. Environmental Science & Policy. 2019;**93**:101-111. DOI: 10.1016/j.envsci.2018.12.033

[27] Pauleit S, Zölch T, Hansen R, Randrup TB, Konijnendijk van den

Bosch C. Nature-based solutions and climate change – Four shades of green. In: Nature-Based Solutions to Climate Change Adaptation in Urban Areas: Linkages between Science, Policy and Practice. Cham, Switzerland: Springer; 2017. pp. 29-49. DOI: 10.1007/978-3-319-56091-5_3

[28] Yazar M, York A. Disentangling justice as recognition through public support for local climate adaptation policies: Insights from the southwest US. Urban Climate. 2022;**41**. DOI: 10.1016/j.uclim.2021.101079

[29] Esperon-Rodriguez M, Rymer P, Power S, Barton D, Carinanos P, Dobbs C, et al. Assessing climate risk to support urban forests in a changing climate. Plants, People, Planet. 2022;**4**(3):201-213. DOI: 10.1002/ppp3.10240

[30] Chrysoulakis N, Somarakis G, Stagakis S, Mitraka Z, Wong MS, Ho HC. Monitoring and evaluating nature-based solutions implementation in urban areas by means of earth observation. Remote Sensing. 2021;**13**(8). DOI: 10.3390/rs13081503

[31] Ascenso A, Augusto B, Silveira C. Rafael Sandraand Coelho S, Monteiro a, Ferreira J, et al. impacts of nature-based solutions on the urban atmospheric environment: A case study for Eindhoven, the Netherlands. Urban Forestry & Urban Greening. 2021;**57**. DOI: 10.1016/j.ufug.2020.126870

[32] Badura T, Lorencova EK, Ferrini S, Vackarova D. Public support for urban climate adaptation policy through nature-based solutions in Prague. Landscape and Urban Planning. 2021;**215**. DOI: 10.1016/j.landurbplan.2021.104215

[33] Davies C, Lafortezza R. Transitional path to the adoption of nature-based solutions. Land Use Policy. 2019;**80**:406-409. DOI: 10.1016/j.landusepol.2018.09.020

[34] Vo TT, Nichersu A, Wendel J. Modeling, monitoring, and validating green roof and green facade solutions with Semantic City models using low cost sensors and open software infrastructures. Urban Science. 2019;**3**(2). DOI: 10.3390/urbansci3020039

[35] Ottelé M, Perini K, Fraaij ALA, Haas EM, Raiteri R. Comparative life cycle analysis for green façades and living wall systems. Energy and Buildings. 2011;**43**(12):3419-3429. DOI: 10.1016/j.enbuild.2011.09.010

[36] Cortinovis C, Geneletti D. Ecosystem services in urban plans: What is there, and what is still needed for better decisions. Land Use Policy. 2018;**70**:298-312. DOI: 10.1016/j.landusepol.2017.10.017

[37] Lehmann S. Growing biodiverse urban futures: Renaturalization and rewilding as strategies to strengthen urban resilience. Sustainability. 2021;**13**(5). DOI: 10.3390/su13052932

[38] Lehmann S. Nature in the urban context: Renaturalisation as an important dimension of urban resilience and planning. Módulo Arquitectura Cuc. 2021;**26**:161-190. DOI: 10.17981/10.17981/MOD.ARQ.CUC.26.1.2021.07

[39] Carvalho R. Introdução ao Direito do Urbanismo. Porto: UCEditora; 2021. pp. 1-269

[40] Correia FA, Correia JA. Regime Jurídico dos Programas e dos Planos Territoriais. Coimbra: Almedina; 2021. pp. 1-494

[41] Hansen R, Olafsson AS, van der Jagt APN, Rall E, Pauleit S. Planning multifunctional green infrastructure for compact cities: What is the state of practice? Ecological Indicators. 2019;**96**:99-110. DOI: 10.1016/j.ecolind.2017.09.042

[42] Van Oijstaeijen W, Van Passel S, Cools J. Urban green infrastructure: A review on valuation toolkits from an urban planning perspective. Journal of Environmental Management. 2020;**267**:110603. DOI: 10.1016/j.jenvman.2020.110603

[43] Haaland C, van den Bosch CK. Challenges and strategies for urban green-space planning in cities undergoing densification: A review. Urban Forestry & Urban Greening. 2015;**14**(4):760-771. DOI: 10.1016/j.ufug.2015.07.009

[44] Beceiro P, Brito RS, Galvao A. Assessment of the contribution of nature-based solutions (NBS) to urban resilience: Application to the case study of Porto. Ecological Engineering. 2022;**175**. DOI: 10.1016/j.ecoleng.2021.106489

[45] Nations U. Resolution adopted by the United Nations Environment Assembly on 2 March 2022. 2022

[46] Agarwal DS, Bharat A. Nature-based solutions for flood–drought mitigation using a composite framework: A case-based approach. Journal of Water and Climate Change. [Internet]. 2023;**14**(3):778-795. Available from: https://www.scopus.com/inward/record.uri?eid=2-s2.0-85152630496&doi=10.2166%2fwcc.2023.369&partnerID=40&md5=ed57cf73a7598e41b9b2b220f83aa591. DOI: 10.2166/wcc.2023.369

[47] O'Donnell E, Thorne C, Ahilan S, Arthur S, Birkinshaw S, Butler D, et al. The blue-green path to urban flood resilience. Blue-Green Systems. 2020;**2**(1):28-45. DOI: 10.2166/bgs.2019.199

[48] Alves A, Vojinovic Z, Kapelan Z, Gersonius SA, B. Exploring trade-offs among the multiple benefits of green-blue-grey infrastructure for urban flood mitigation. Science of the Total Environment. 2020;**703**. DOI: 10.1016/j.scitotenv.2019.134980

[49] Bark RH. Designing a flood storage option on agricultural land: What can flood risk managers learn from drought management? Water (Switzerland). [Internet]. 2021;**13**(18). Available from: https://www.scopus.com/inward/record.uri?eid=2-s2.0-85115420774&doi=10.3390%2fw13182604&partnerID=40&md5=67eda19434f3ebc7dc7c49c9678368d2. DOI: 10.3390/w13182604

[50] Bernello G, Mondino E, Bortolini L. People's perception of nature-based solutions for flood mitigation: The case of Veneto region (Italy). Sustainability. 2022;**14**(8). DOI: 10.3390/su14084621

[51] Chan FKS, Griffiths JA, Higgitt D, Xu S, Zhu F, Tang YT, et al. "Sponge City" in China-a breakthrough of planning and flood risk management in the urban context. Land Use Policy. 2018;**76**:772-778. DOI: 10.1016/j.landusepol.2018.03.005

[52] Connelly A, Snow A, Carter J, Wendler J, Lauwerijssen R, Glentworth J, et al. What approaches exist to evaluate the effectiveness of UK-relevant natural flood management measures? A systematic map. Environmental Evidence. [Internet]. 2023;**12**(1). Available from: https://www.scopus.com/inward/record.uri?eid=2-s2.0-85160028571&doi=10.1186%2fs13750-023-00297-z&partnerID=40&md5=e5f94ee0c83e30dc61a8824b7481df3b. DOI: 10.1186/s13750-023-00297-z

[53] Davies HJ, Doick KJ, Hudson MD, Schaafsma M, Schreckenberg K, Valatin G. Business attitudes towards funding ecosystem services provided by urban forests. Ecosystem Services. 2018;**32**(B):159-169. DOI: 10.1016/j.ecoser.2018.07.006

[54] Turhan C, Atalay AS, Akkurt GG. An integrated decision-making framework for mitigating the impact of urban Heat Islands on energy consumption and thermal comfort of residential buildings. Sustainability. 2023;**15**(12). DOI: 10.3390/su15129674

[55] Bosch M, Locatelli M, Hamel P, Remme RP, Jaligot R, Chenal J, et al. Evaluating urban greening scenarios for urban heat mitigation: A spatially explicit approach. Royal Society Open Science. 2021;**8**(12). DOI: 10.1098/rsos.202174

[56] Marando F, Heris MP, Zulian G, Udías A, Mentaschi L, Chrysoulakis N, et al. Urban heat island mitigation by green infrastructure in European functional urban areas. Sustainable Cities and Society. [Internet]. 2022;**77**:103564. Available from: https://www.sciencedirect.com/science/article/pii/S2210670721008301. DOI: 10.1016/j.scs.2021.103564

[57] Worku H. Integrating climate change adaptation strategies in urban planning and landscape design of Addis Ababa City, Ethiopia: Using urban planning and landscape design to mitigate flooding, drought, and urban heat island effects. Environmental Quality Management. [Internet]. 2017;**27**(1):5-21. Available from: https://www.scopus.com/inward/record.uri?eid=2-s2.0-85040124677&doi=10.1002%2ftqem.21514&partnerID=40&md5=7ed284fcce88d89acd622613c0bea740

[58] Nastran M, Kobal M, Eler K. Urban heat islands in relation to green land use in European cities. Urban Forestry & Urban Greening. 2019;**37**:33-41. DOI: 10.1016/j.ufug.2018.01.008

[59] Anderson V, Gough WA, Agic B. Nature-based equity: An assessment of the public health impacts of green infrastructure in Ontario Canada. International Journal of Environmental Research and Public Health. 2021;**18**(11). DOI: 10.3390/ijerph18115763

[60] Pearce H, Levine JG, Cai X, MacKenzie AR. Introducing the green infrastructure for roadside air quality (GI4RAQ) platform: Estimating site-specific changes in the dispersion of vehicular pollution close to source. Forests. 2021;**12**(6). DOI: 10.3390/f12060769

[61] del Redondo Bermudez MC, Miguel Kanai J, Astbury J, Fabio V, Jorgensen A. Green fences for Buenos Aires: Implementing green infrastructure for (more than) air quality. Sustainability. 2022;**14**(7). DOI: 10.3390/su14074129

[62] Sebastiani A, Buonocore E, Franzese PP, Riccio A, Chianese E, Nardella L, et al. Modeling air quality regulation by green infrastructure in a Mediterranean coastal urban area: The removal of PM10 in the Metropolitan City of Naples (Italy). Ecological Modelling. 2021;**440**. DOI: 10.1016/j.ecolmodel.2020.109383

[63] Castelo S, Amado M, Ferreira F. Challenges and opportunities in the use of nature-based solutions for urban adaptation. Sustainability. 2023;**15**(9):7243. DOI: 10.3390/su15097243

[64] Brom P, Engemann K, Breed C, Pasgaard M, Onaolapo T, Svenning JC. A decision support tool for green infrastructure planning in the face of rapid urbanization. Land.

2023;**12**(2):415. DOI: 10.3390/land12020415

[65] Almenar JB, Elliot T, Rugani B, Philippe B, Gutierrez TN, Geneletti SG, et al. Nexus between nature-based solutions, ecosystem services and urban challenges. Land Use Policy. 2021;**100**. DOI: 10.1016/j.landusepol.2020.104898

[66] Boros J, Mahmoud I. Urban Design and the role of Placemaking in mainstreaming nature-based solutions. Learning from the Biblioteca Degli Alberi case study in Milan. Frontiers in Sustainable Cities. 2021;**3**. DOI: 10.3389/frsc.2021.635610

[67] Kabisch N, Frantzeskaki N, Pauleit S, Naumann S, Davis M, Artmann M, et al. Nature-based solutions to climate change mitigation and adaptation in urban areas: Perspectives on indicators, knowledge gaps, barriers, and opportunities for action. Ecology and Society. 2016;**21**(2). DOI: 10.5751/ES-08373-210239

[68] Commission E. Communication from the Commission to the European Parliament, The Council, The European Economic and Social Committee and the Committee of the Regions - Green Infrastructure (GI) — Enhancing Europe's Natural Capital COM/2013/0249 Final. United States: Oxford Academic; 2013

[69] Esmail BA, Cortinovis C, Suleiman L, Albert C, Geneletti D, Mörtberg U. Greening cities through urban planning: A literature review on the uptake of concepts and methods in Stockholm. Urban Forestry & Urban Greening. 2022;**72**. DOI: 10.1016/j.ufug.2022.127584

[70] Commission E. Communication from the Commission to the European Parliament, The Council, The European Economic and Social Committee and The Committee of the Regions - EU Biodiversity Strategy for 2030 - COM(2020) 380 Final. [Internet]. 2020. Available from: https://eur-lex.europa.eu/legal-content/EN/TXT/?uri=celex%3A52020DC0380

[71] Yuan J, Wu B, Liu X, Lu M. Boundary green infrastructure: A green infrastructure connecting natural and artificial spaces. Frontiers in Environmental Science. 2023;**11**. DOI: 10.3389/fenvs.2023.1155036

Chapter 8

# Urban Cultural Heritage under Threat: Strategies for Sustainable Use and Modernisation in Non-EU Contexts

*Sérgio Lousada, Andrii Khorolskyi, Myroslava Lendel, Oksana Sviezhentseva and Svitlana Delehan*

**Abstract**

This chapter investigates the challenges and solutions related to the energy-efficient modernisation of a cultural heritage site without compromising its historical value. The problem addressed involves the low energy performance of historical buildings in urban environments, particularly in non-EU countries facing infrastructural constraints and security risks. The chapter aims to present a replicable approach for enhancing the sustainability of heritage architecture through technically and environmentally sound interventions. A case study was conducted on a building constructed in 1885, currently housing the Educational and Research Institute of Chemistry and Ecology at Uzhhorod National University. The methodology involved a comprehensive energy audit, an assessment of urban density changes (2017–2024), and a simulation-based evaluation of retrofitting measures. Proposed interventions included ceiling insulation, window and door replacement, installation of an individual heating unit, thermostatic and balancing valves, and a mechanical ventilation system. As a result, total energy savings of 503,220 kWh/year and a $CO_2$ reduction of 111.5 tonnes/year were achieved. The building's energy efficiency class improved from "C" to "B". This chapter contributes to the field by offering a structured model for sustainable modernisation applicable to heritage buildings in regions undergoing rapid transformation and lacking unified energy policy frameworks.

**Keywords:** modernisation, cultural heritage, energy consumption, $CO_2$ emissions, sustainable development, Ukraine

## 1. Introduction

The preservation and energy-efficient modernisation of cultural heritage buildings present a pressing challenge in the context of urban sustainability, especially in non-EU countries facing war-induced infrastructural strain and limited policy integration. This chapter explores the intersection of cultural value preservation and the need for thermal performance improvement in historic buildings, using the case of Uzhhorod as an example. The research is guided by the following question: How can energy efficiency be enhanced in heritage buildings without compromising their historical integrity, particularly in cities experiencing high urban density and geopolitical risks?

The chapter is structured as follows: Section 1.1 discusses the broader European and Ukrainian perspectives on cultural heritage and energy efficiency. Section 1.2 addresses the legal and institutional frameworks in Ukraine, with emphasis on post-2022 developments. Subsequent sections present a detailed case study, including energy audit results, technical assessments, proposed interventions, and outcome evaluation in terms of energy savings and $CO_2$ emissions. By presenting a replicable model of heritage building retrofitting in a high-risk and resource-constrained environment, this chapter contributes to bridging the gap between preservation ethics and energy resilience in urban planning.

### 1.1 Preservation and modernisation of cultural heritage sites in the context of urbanisation

The preservation and modernisation of cultural heritage sites in urbanised environments represent a critical challenge for contemporary urban planning and sustainable development. The European Union recognises cultural heritage as an essential component of identity, memory, and economic growth, integrating it into strategic policies related to environmental sustainability, energy efficiency, and risk management [1]. In EU countries, substantial efforts are made to enhance the resilience and energy performance of historic buildings, whereas in non-EU countries, similar approaches are often neglected, leading to excessive energy losses and inefficient utilisation of cultural assets [2].

The European Cultural Convention, which Ukraine joined in 1994, emphasises the importance of protecting national contributions to European heritage [3]. However, urbanised environments pose unique challenges to the conservation of historic buildings due to high population density, infrastructure stress, and inadequate adaptation to modern energy efficiency standards [4]. Addressing these challenges requires a balanced approach that maintains architectural authenticity while incorporating modern energy solutions [5].

The European Union has developed integrated policies for heritage conservation, emphasising sustainability and climate adaptation. Initiatives such as the European Green Deal and the New European Bauhaus encourage the adaptive reuse of historic buildings, ensuring compliance with energy performance standards while maintaining their cultural significance [6]. The Heritage Adaptation and Energy Efficiency Framework highlights the necessity of energy-efficient solutions tailored to historic structures, incorporating passive design strategies, renewable energy integration, and intelligent climate control systems [7].

The protection of cultural heritage within the EU is reinforced through several sectoral policies, including risk management frameworks, which address emergency preparedness for heritage sites [8], sustainability action plans, which integrate conservation with carbon footprint reduction [9], and legal mechanisms against the illicit trade and degradation of cultural assets [10]. Despite these advancements, historic buildings in many non-EU countries remain underfunded and lack systematic integration of energy-efficient modernisation [11]. The absence of financial incentives and legal frameworks often results in energy waste and the gradual deterioration of cultural sites [12].

Uzhhorod, a historic city in western Ukraine, presents a unique case of cultural heritage management in a non-EU context. With a history spanning over 1100 years, the city retains significant architectural assets, including medieval, Austro-Hungarian, and Czech constructivist influences [13]. However, contemporary urbanisation, demographic shifts and displacement due to war have increased stress on its historic infrastructure, necessitating urgent modernisation efforts [14].

The period of independent Ukraine in the history of Uzhhorod can be briefly characterised by the following key social processes that influenced the urban development and shaping of the historic city centre:

1. The economic crisis is one of the most significant factors that affected the state of the urban landscape, leading to the decline of industrial enterprises, housing and communal facilities, turning some neighbourhoods into ruins and wastelands.

2. The lifting of the moratorium on the construction of individual housing has led to a significant increase in urban development, especially in the form of individual housing blocks, while losing historic landscapes.

3. The privatisation of housing and the decline of the public utilities sector have worsened the condition of residential buildings, affecting the roofs, facades and interior spaces of neighbourhoods.

4. The destruction of historical elements and disregard for architectural principles have led to a loss of historical value and aesthetics of the city centre.

5. The spontaneous reconstruction of historic buildings, despite attempts to regulate it, has led to chaos and misunderstandings in the city's development.

6. Lack of enforcement of urban and monument protection legislation has led to chaotic development and destruction of the authentic environment of the historic centre.

**Figure 1** shows 30 architectural monuments of Uzhhorod Malyi Galagov that could be included in the UNESCO list of monuments:

1. Masaryk Bridge.

2. Rafanda.

3. Zemstvo gendarmerie command of Subcarpathian Rus.

4. House of public health.

5. Zemstvo government of Subcarpathian Rus.

6. Justice block.

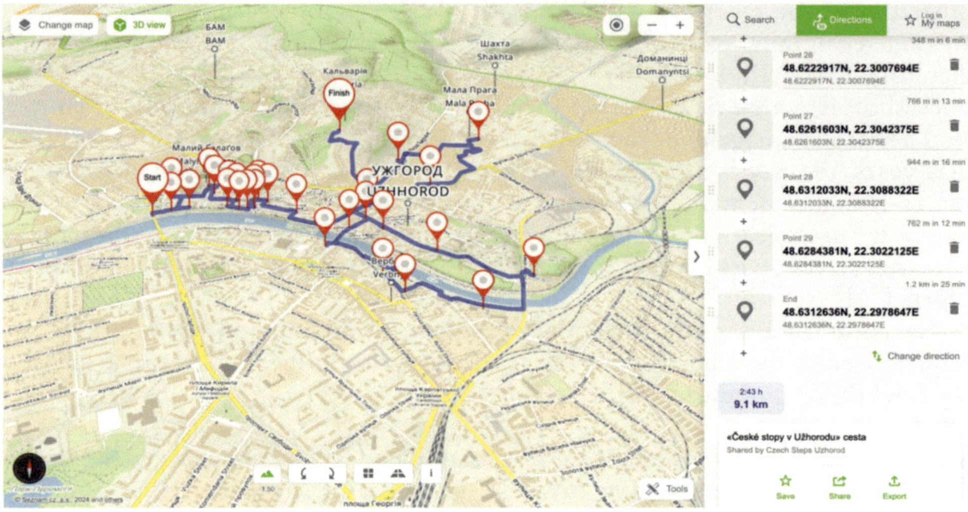


**Figure 1.**
*Objects of the Austro-Hungarian heritage in Malyi Galagov district (Uzhhorod).*

7. Residential buildings of Malyi Galagov.
8. House for military personnel.
9. Masaryk school.
10. Linden alley.
11. National bank.
12. Zemstvo financial administration of Subcarpathian Rus.
13. Command of the 12th Czechoslovak Infantry Division.
14. Temporary government building.
15. State Police Department of Subcarpathian Ruthenia.
16. Postal and telegraphic government.
17. The building of the Legio cooperative.
18. City cinema and library.
19. Baty Palace.
20. House of Dr. Shalamon.
21. People's House of the O. Dukhnovich Society.
22. Craft school.

23. Hebrew-speaking gymnasium.

24. City swimming pool.

25. Ivan Olbracht Street.

26. Jewish school and community centre.

27. People's House of the Prosvita Society.

28. Lesser Prague.

29. Villa of Antony Beskyd.

30. Municipal water supply building and hydrant network.

The city of Uzhhorod in Ukraine has the richest architectural heritage of the Czechoslovak interwar modernism era, and local urbanists and cultural heritage advocates dream of adding the Malyi Galagov district to the UNESCO list of monuments. But key challenges are not only the necessity of modernisation, which include outdated building materials contributing to high energy consumption [15]; insufficient insulation and heating systems that fail to meet modern efficiency standards [16]; increased occupancy and urban stress due to the relocation of displaced persons since 2022 [17]; lack of regulatory enforcement for energy-efficient retrofits in heritage structures [18].

Comparative studies on heritage preservation in EU cities such as Košice (Slovakia), Nyíregyháza (Hungary), and Rzeszów (Poland) indicate that systematic policy implementation and financial support mechanisms significantly enhance the longevity and efficiency of historic buildings. The adaptation of such approaches to Uzhhorod could provide substantial long-term benefits in terms of sustainability and cultural asset conservation [19].

This study investigates the modernisation of cultural heritage sites in a densely developed city, focusing on integrating European best practices for sustainability and energy efficiency. By addressing these objectives, this research contributes to the broader discussion on sustainable urban development and heritage conservation in non-EU cities facing modernisation challenges.

### 1.2 Analysis of the experience of implementing the Hague convention for the protection of cultural property in the event of armed conflict in Ukraine

The best practices for implementing the provisions of the Hague Convention for the Protection of Cultural Property in the Event of Armed Conflict (1954) and its Protocols (1954 and 1999), as identified through the analysis of national reports for the period 2017–2020, reflect a multifaceted approach to cultural heritage protection. These practices recognise that cultural property can hold varying degrees of significance—local, national, or international. Moreover, it may exist in diverse forms: tangible or intangible, movable or immovable, and under both public and private ownership.

To ensure comprehensive documentation and protection, many countries promote the development of private asset inventories, providing advocacy and advisory support to stakeholders. As a common strategy, inventories are organised by categories that include intangible heritage, tangible heritage, urban heritage, and natural heritage [20]. The creation of an accurate inventory of cultural property contributes to several aspects, including:

- Combating theft and trafficking of cultural property or facilitating its return in case of looting.

- Assisting armed forces in identifying protected heritage and including it in lists of prohibited objects during conflict.

- Providing sufficient and accurate data for reconstruction needs before the destruction of the site [21].

The mapping of inventoried cultural property serves to inform civilian and military personnel about objects that require special attention in the event of an emergency, for example, by including them in lists of objects prohibited for shelling or marking them on military maps. A Geographic Information System (GIS) database is often used for this purpose [22]. The creation of a publicly accessible national online database is a common practice for compiling national inventories. Inventories can be used by national security forces for operational planning and deployment in the field [23]. The use of digital technologies, such as digitisation, allows cultural property to be protected even in the event of physical destruction. For example, the use of 3D technologies provides an opportunity for research and preservation of endangered monuments. A good practice is to make an online database of digitised cultural property available [24]. The digitisation of cultural property at the national level as part of the UNESCO Memory of the World programme helps to attract financial resources for the preservation of cultural heritage. In addition, digital technologies can be applied to private collections and antiquities that have disappeared or been stolen [25].

National and local emergency plans provide for risk assessment and the development of strategies for the protection of cultural property in the event of conflicts. There is a practice of establishing a national body that coordinates measures to protect cultural property in emergency situations [26]. Legislative acts are an integral part of the legal framework governing the study of historic cities in Ukraine. Due to the unique nature and significance of these cities, legislation defines the basic principles of their development, reconstruction, and protection of cultural heritage. Legal norms such as: Law of Ukraine "On the Protection of Cultural Heritage", Law of Ukraine "On Architectural Activity", Law of Ukraine "On Investment Activity", and Law of Ukraine "On Regulation of Urban Development Activities" establish rules for the use and protection of historic areas [27–31]. The concept of "cultural heritage" is defined as a set of cultural heritage objects inherited from previous generations of people. However, the law does not define 'cultural heritage of the Ukrainian people' and does not take into account the importance of cultural heritage for the Ukrainian people and the protection of intangible cultural heritage [32–34]. Thus, the basic document on the protection of cultural heritage is not in line with the UNESCO Convention for the Safeguarding of the Intangible Cultural Heritage and the Council of Europe Framework Convention on the Importance of Cultural Heritage for Society [35, 36].

One of the key tasks is to develop a historical and architectural reference plan for the city of Uzhhorod, which includes the definition of monument protection zones, historical areas, and an inventory of cultural heritage. These processes are based on a number of legal documents that form the current regulatory framework for research in the field of protection and development of historic cities [37]. To ensure the sustainable future of these areas, such planning must also integrate guidelines for the energy modernisation of historic buildings, aligned with national and European sustainability goals. Incorporating circular economy principles at this stage would enable not only the conservation of architectural value but also the adaptive reuse of resources and energy efficiency.

## 2. Study of the changing landscape of the city

This section aims to present the methodology for assessing urbanisation dynamics and outlines the sequence and rationale for selecting modernisation measures for a historical, cultural heritage site under pressure from growing urban density.

At the initial stage of this chapter, an assessment of changes in the urban landscape was conducted through a comparative analysis of satellite images from 2017 and 2024. This analysis was facilitated by the EO Browser, a cloud-based platform that enables the visualisation and retrieval of medium- and low-resolution satellite imagery, including datasets from Sentinel-1, Sentinel-2, Sentinel-3, Sentinel-5P, Landsat, MODIS, and others. The tool integrates multiple functionalities such as automatic visualisations (e.g., true colour and NDVI), spectral band composition, statistical analysis, and custom script-based data processing.

To illustrate the level of urban development, satellite imagery from Sentinel-2A for a cloud-free spring date—5 May 2024—was selected and classified using the Urban Classified Script. For comparison, a similar image from 2 April 2017 was processed. A "Green City" composite image for summer 2024 was also generated to visualise vegetation distribution [38].

The EO Browser enabled instant access to and rendering of satellite mosaics through automated selection, download, and pre-processing of suitable image scenes. To define the area of interest, the city of Uzhhorod was selected through the search interface, with imaging parameters set to Landsat 8 L1, maximum 10% cloud coverage, and time range adjusted accordingly. The selected image was visualised using the true colour option.

Subsequently, the same procedure was replicated for the 2017 dataset to ensure consistency. By switching to Education Mode, a simplified interface and thematic filtering option became available. The Urban theme was selected to allow classification using Sentinel-2 imagery. Within the defined time frame (1–5 May 2024), a cloud-free image was visualised and colour-coded using the Urban Classified scheme, which distinguishes four land cover categories: white for built-up areas, blue for water bodies, brown for bare land, and green for vegetation (**Figures 2** and **3**) [39].

The Urban Classified scenario is applied to distinguish built-up urban zones from surrounding barren land, vegetation, and water bodies. Within this classification, moisture-rich areas are rendered in blue, urbanised territories in white, vegetated surfaces in green, and non-vegetated open ground in brown, thereby enabling a clearer spatial interpretation of urban structure.

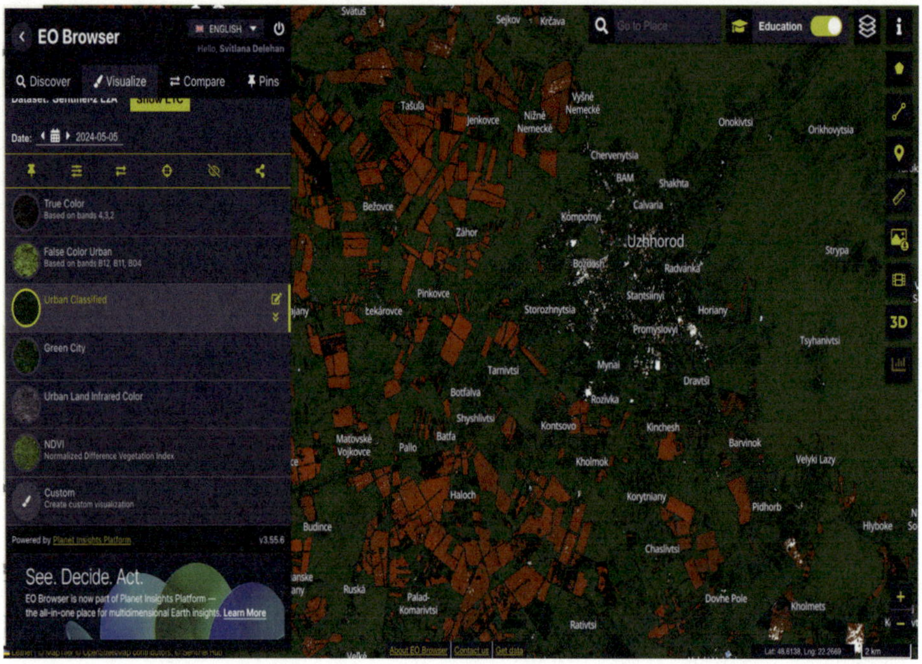


**Figure 2.**
*Territory of the city Uzhhorod 05/05/2024 (Urban classified via EO browser).*

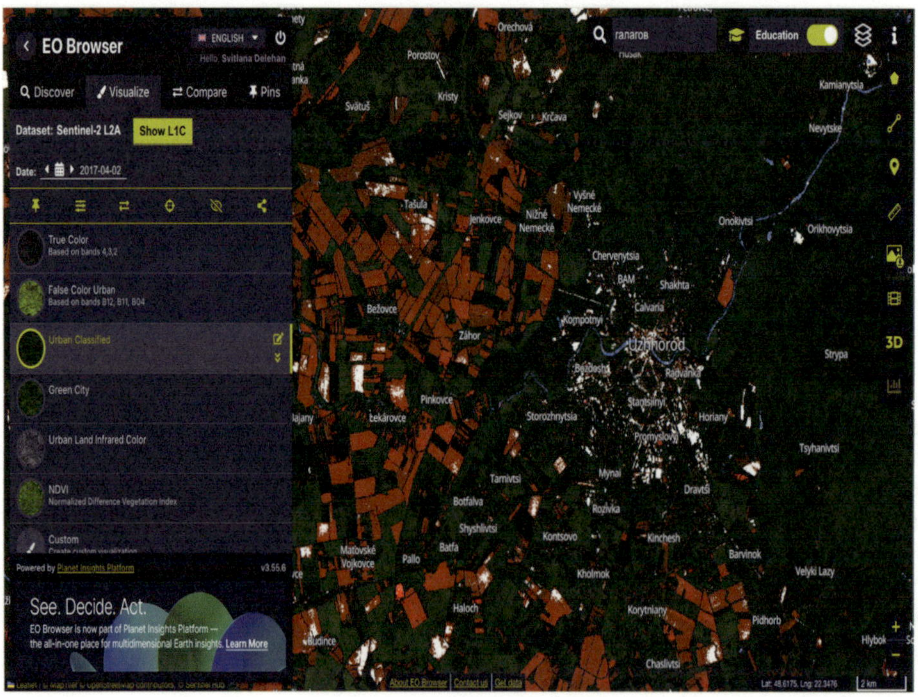


**Figure 3.**
*Territory of the city Uzhhorod 02/04/2017 (Urban classified via EO browser).*

*DOI: http://dx.doi.org/10.5772/intechopen.1010319*

The EO Browser's educational mode was utilised to examine the extent of green space within the study area. From the Highlights section under the Urban theme, a pre-processed image of the selected city was accessed from the platform's thematic image library. All library imagery is displayed by default in true colour mode. To assess the spatial distribution and quantity of green zones, the Green City script was applied. This allowed the green areas to be visually identified and delineated directly from the map.

The employed script uses the Normalised Differential Vegetation Index (NDVI) in combination with specific natural colour wavelengths to calculate vegetative cover. As a result, the visualisation distinguishes between built-up areas, which are displayed in grey, and vegetated areas, which appear in green. This classification approach proves effective for comparative analysis of urban greenery in different geographic contexts (**Figures 4** and **5**) [40].

The Normalised Differential Vegetation Index is a simple but effective indicator for quantifying green biomass. It is an indicator of vegetation health based on how plants reflect waves of light of a certain length. The range of NDVI values is from −1 to 1. Negative NDVI values (values close to −1) indicate proximity to water. Values close to zero (from −0.1 to 0.1) usually indicate a location on exposed rocks, sand or snow. Low positive values are typical for shrublands and grasslands (approximately 0.2 to 0.4), while high values are typical for temperate and tropical forest vegetation (values close to 1) (**Figures 6** and **7**).

The Normalised Differential Vegetation Index (NDVI) was applied to assess the state of vegetation in the territory of Uzhhorod as of 05/05/2024, using satellite imagery accessed via EO Browser. The NDVI is a widely used indicator for evaluating

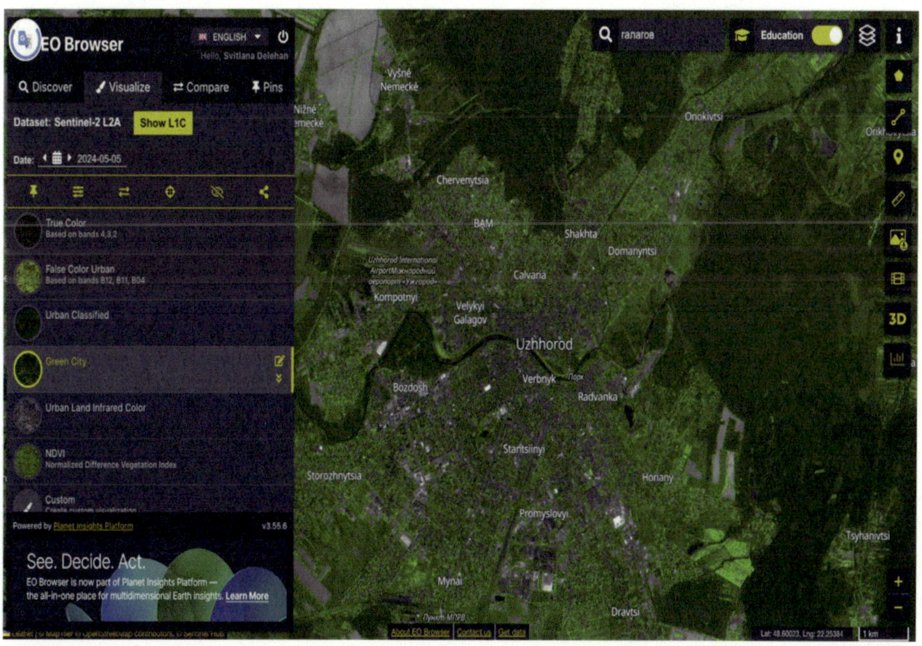


**Figure 4.**
*Territory of the city Uzhhorod 05/05/2024 (Green city via EO browser).*

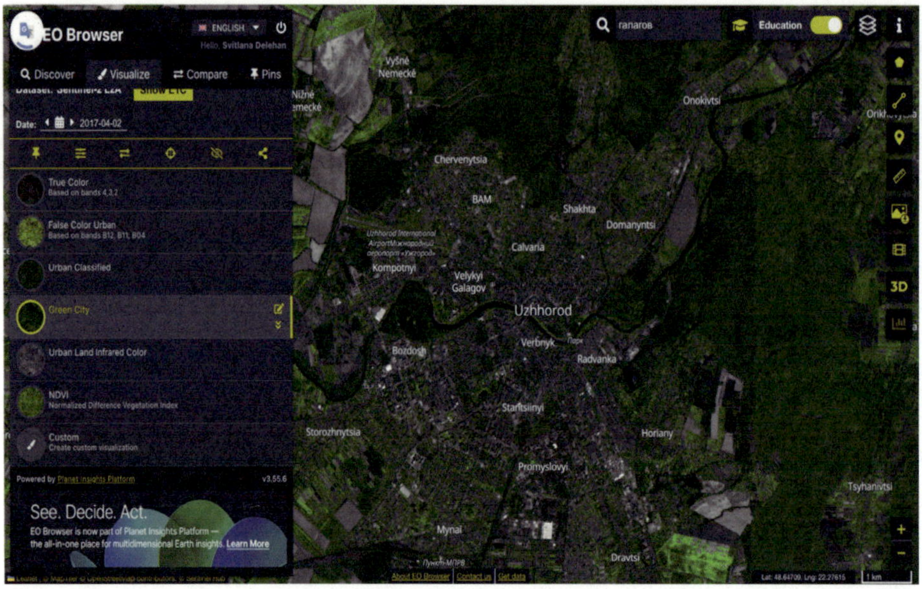


**Figure 5.**
*Territory of the city Uzhhorod 02/04/2017 (Green city via EO browser).*

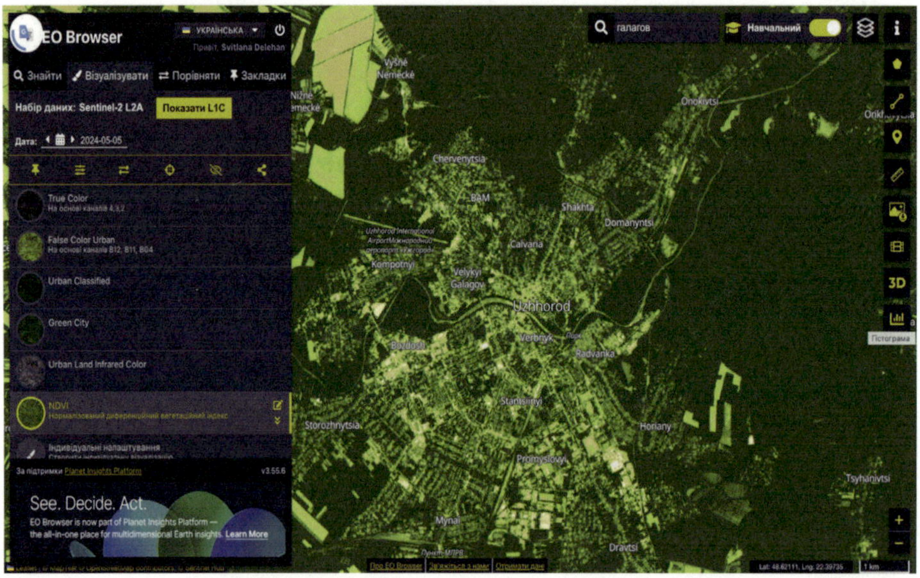


**Figure 6.**
*Territory of the city Uzhhorod 05/05/2024 (NDVI via EO browser).*

plant health, based on the reflectance of specific light wavelengths, and it serves as an effective tool for classifying vegetation types across landscapes.

The NDVI data obtained reveal that negative values (approaching −1) correspond to water bodies located near the analysed territory. Values near zero (between −0.1 and 0.1) are typically associated with non-vegetated surfaces such as rocky areas, sand, or snow. Moderate positive values (from approximately 0.2 to 0.4) indicate the

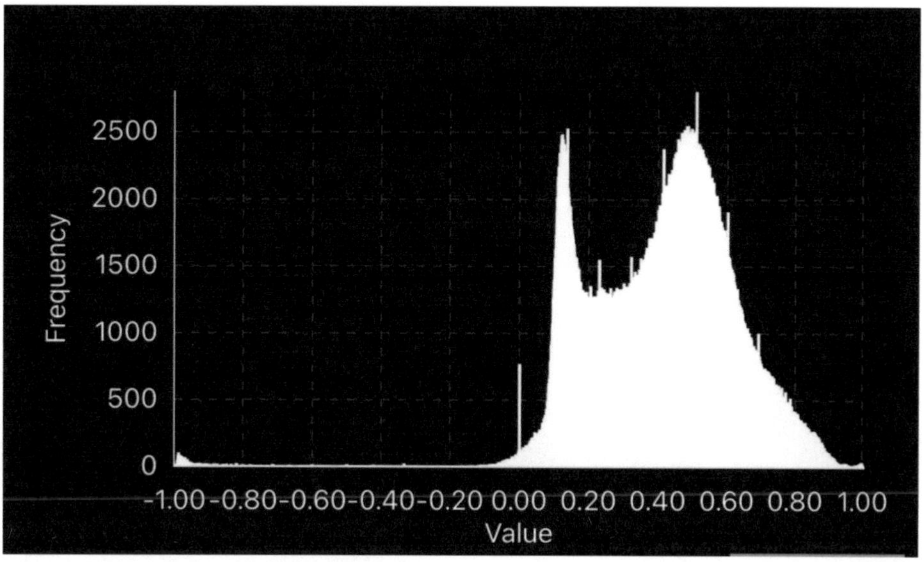


**Figure 7.**
*The normalised differential vegetation index for the territory of the city Uzhhorod 05/05/2024 (via EO browser).*

presence of shrubland and grassland, whereas high values (close to 1) are characteristic of dense forest cover, which is typical of temperate and tropical ecosystems [7, 8].

Furthermore, the temporal profile of NDVI values suggests that the highest activity peaks occur between 01:00 and 08:00, potentially reflecting increased photosynthetic activity during this interval. Overall, the NDVI-based analysis conducted for the selected period provides valuable insights into the composition and health of vegetative cover within Uzhhorod. Such data support environmental monitoring, land use planning, and the sustainable management of natural resources in the context of ongoing urban development (**Figures 8** and **9**).

The description of the Normalised Differential Vegetation Index (NDVI) for the territory of Uzhhorod as of 02/04/2017, carried out using EO Browser, provides important information about the state of vegetation cover in this region.

According to the data obtained, the NDVI values range from −0.04 to 0.85, which indicates the diversity of vegetation types and their condition in the specified area. The highest peaks are observed in the period from 07:00 to 08:00, which may indicate the active photosynthetic activity of plants at this time.

The frequency of the data obtained is 4000, which indicates a high resolution of the information and the ability to detail changes in vegetation cover over time. The information obtained is important for assessing the health and development of vegetation, as well as for monitoring the ecological condition of the territory [41].

The study of landscape change and population dynamics allowed us to make certain generalisations and track development trends, which are as follows:

1. From 2011 to 2024, the housing stock increased from 2514 thousand m2 to 2803 thousand m2. This indicates the active development of the city's construction sector.

2. 74% of the total area is multi-apartment buildings. This means a high population density in the city, which is reflected in the large number of apartment buildings.

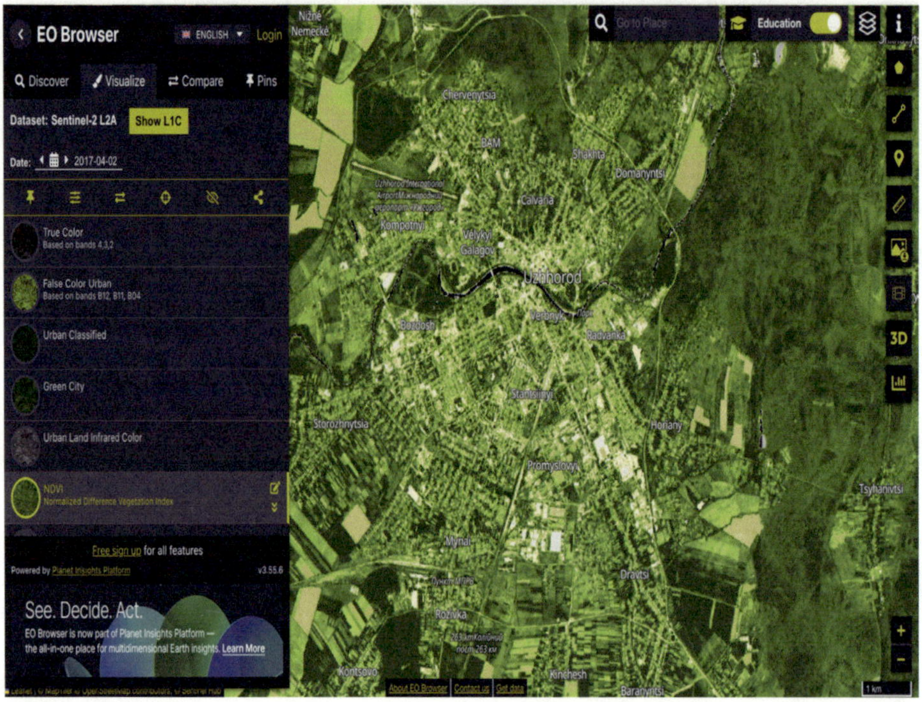


**Figure 8.**
*Territory of the city Uzhhorod 02/04/2017 (NDVI via EO browser).*

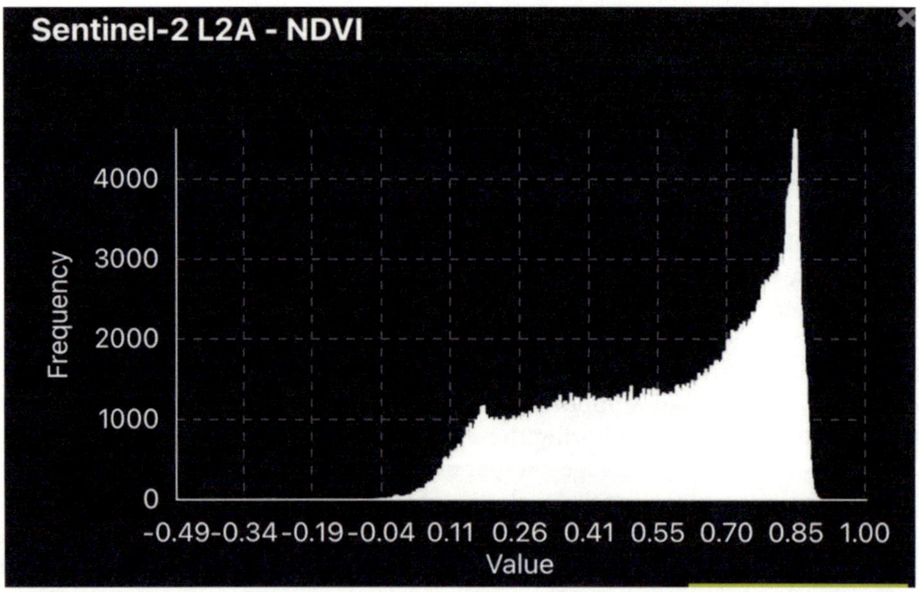


**Figure 9.**
*The normalised differential vegetation index for the territory of the city Uzhhorod 02/04/2017 (via EO browser).*

3. The main goal of the housing component of the city's strategy is a comprehensive modernisation and reconstruction of the existing housing stock. This means that the city focuses its efforts on improving existing buildings and residential facilities.

4. The seismic specificity of the territory limits the possibilities of increasing the density and number of storeys of buildings. This presents the city with complex planning and development challenges.

5. The general condition of the street network and road facilities does not meet modern technical and operational standards. This requires serious investment efforts to improve the condition of roads and infrastructure.

Selection of modernisation measures was based on the results of a comprehensive energy audit and microclimate analysis. The following criteria were used to select suitable measures:

1. Compatibility with historical value and architectural features;

2. Contribution to reducing heat loss and $CO_2$ emissions;

3. Compliance with energy efficiency regulations;

4. Potential to improve indoor comfort and microclimate stability;

5. Feasibility of implementation under current budget and technical constraints.

Based on these criteria, a step-by-step modernisation strategy was developed, including:

1. Thermal insulation of the ceiling and attic structures to reduce heat loss through the upper envelope;

2. Replacement of outdated window and door units with energy-efficient models preserving visual authenticity;

3. Installation of individual heating control points and thermostatic valves for system balancing and consumption optimisation;

4. Implementation of mechanical ventilation systems to ensure air quality while reducing infiltration-related losses;

5. Integration of automatic balancing valves on risers to stabilise heat distribution;

6. Recommendations for lighting optimisation and transition to LED solutions.

A flowchart of this decision-making process and implementation sequence is recommended for future publication versions to enhance readability and reproducibility.

The combination of energy efficiency upgrades and preservation of architectural integrity demonstrates the potential for circular economy principles to be effectively applied to cultural heritage buildings. Moreover, the experience gained from this case study supports the feasibility of scaling similar interventions to other historic buildings in urbanised regions facing environmental and geopolitical challenges. Future research may explore: the long-term monitoring of energy performance post-modernisation; adaptive reuse strategies incorporating renewable energy; development of legal and financial frameworks for non-EU countries aiming to align with EU sustainability targets.

## 3. Results of research on the modernisation of a cultural heritage site

The growing relevance of sustainable urban planning and energy-efficient retrofitting, especially in the context of cultural heritage preservation, has led to a case study of the energy retrofitting of a specific cultural heritage site in Uzhhorod, which serves as both a representative example of the city's architectural heritage and a practical demonstration of sustainable renovation approaches. The following section presents the results of this in-depth study, with a detailed description of the selected site, its historical and technical characteristics, and the key areas of the proposed modernisation.

### 3.1 Description of the facility, climatic conditions, areas of modernisation

The energy modernisation of cultural heritage sites is an important and interesting topic, especially relevant for Uzhhorod, which is an ancient Ukrainian city and could potentially be included in the UNESCO World Heritage List. The city, which is described as a national masterpiece of art, is facing the need to transform its buildings and cultural heritage sites to improve their energy efficiency, aimed at sustainable development of the city. This means optimising the use of resources, including water and energy, and reducing $CO_2$ emissions [42, 43].

Uzhhorod has a rich history and cultural heritage, which includes more than 2000 monuments. However, many of them demonstrate low levels of energy efficiency, resulting in significant $CO_2$ emissions and high energy bills. In addition, many of these structures, which have existed for over 200 years, have become unsafe to operate due to natural ageing, urbanisation, changing hydrogeological conditions and other factors.

Uzhhorod also has problems with its engineering infrastructure, which is in poor technical condition. This indicates the need for energy modernisation of cultural heritage sites. Special institutions, such as the Department for the Protection of the Historic Environment, are being created around the world to help solve this problem. There are different approaches to the energy rehabilitation of cultural heritage sites, and it is important to find the best way for Uzhhorod.

Before starting renovation, it is necessary to analyse the energy-saving potential and opportunities in historic buildings. This analysis will help determine which measures will be effective from an energy point of view and which require further study. In historic buildings, air conditioning and heating systems are often in poor condition and require regular maintenance, while meeting environmental and economic requirements. An important aspect of energy modernisation is the preservation of architectural values and unique artefacts, so energy-efficient materials and technologies should be used.

The reconstruction and thermal modernisation of walls, especially in historic buildings, is a complex problem. The use of internal wall insulation can be an alternative to preserve architectural value, but it requires new technologies and approaches to thermal diagnostics. The assessment of historic buildings developed by the Italian Green Building Council aims to preserve architectural heritage and improve energy efficiency.

Managing the internal air in historic buildings is also important for improving energy efficiency. Using internal air ducts to remove moisture can help improve the quality of the walls. Energy efficiency issues are also related to the impact of climate change, which can lead to flooding and other problems in historic cities [44, 45].

Despite the challenges, energy retrofitting of historic buildings is important for improving the quality of life in cities. Although there are challenges, such as the limited use of modern materials and technologies, compromise solutions need to be found to preserve the unique character of historic buildings and at the same time improve their energy efficiency [46–48].

The building of the Educational and Research Institute of Chemistry and Ecology of Uzhhorod National University is considered as a modernisation object. The building is located at 53 Fedynets Street, Uzhhorod. This facility is owned by the institution and operates as a non-profit organisation. The construction of the building was completed in 1855. The premises are heated by a district heating system. However, there is no hot water supply.

As for electricity, it is supplied through the regional power grid. Cold water is supplied from the central water supply system, and wastewater is discharged into the sewerage system. The building has a frameless construction with load-bearing brick walls. There is also a heated basement. The upper floor of the building is a cold attic. The building has the status of an architectural monument of local significance.

Since its construction, the building has undergone regular cosmetic repairs. These repairs have included upgrading the lighting in the premises and repairing the roof. **Table 1** provides a description of the building.

To assess the effect of modernisation, we analysed changes in utility costs. **Table 2** provides information on service providers and service prices.

The city of Uzhhorod is located in the II climatic zone of Ukraine, according to DBN B.2.6-31:2016. According to DSTU-N B V.1.1-27-2010 [49, 50]. Uzhhorod belongs to the Transcarpathian architectural and construction district (Sub-district IIIB). The average relative humidity of the outside air is 72%. The prevailing winds during the heating season are southeast. The annual rainfall is 745 mm. **Table 3** shows the heating data for the building.

After that, an energy audit of the building was carried out. The following Standards and Regulations are in force in Ukraine, which are used to guide the energy performance of buildings [51–69].

### 3.2 Building energy audit data

The energy audit of the selected cultural heritage site was conducted to assess the current state of energy consumption and the efficiency of existing engineering systems, with the goal of identifying potential areas for improvement. This process is essential for planning energy-efficient modernisation measures, particularly in historic buildings that require both preservation and adaptation to contemporary comfort and sustainability standards. In accordance with the terms of the audit agreement, the primary objectives included the inspection of the building envelope

| **Building characteristics** | | | | | |
|---|---|---|---|---|---|
| **Name** | **Educational and research institute of chemistry and ecology** | | | | |
| Assignment | Educational institution | | | | |
| Average number of students | 249 | | | | |
| Average number of staff | 70 | | | | |
| Year of commissioning | 1855 | | | | |
| Location | Detached building | | | | |
| An architectural monument | Yes | | | | |
| Number of units, pcs. | The total area, м$^2$ | Total volume, м$^3$ | Condition. area, м$^2$ | Condition. volume, м$^3$ | |
| 3 | 4789,4 | 16,872,1 | 4782,4 | 16,849,7 | |
| Heating operation mode and schedule of people's stay | | | | | |
| Schedules | | Working days | Saturday | Sunday | Holidays days |
| Schedule of people's stay, hours/day | | 10 | — | — | — |
| Heating schedule, h/day | | 24 | | 24 | |

**Table 1.**
*Building characteristics (before modernisation).*

| Name of utility services | Name of the company providing the service | *Unit price (including VAT) |
|---|---|---|
| Supply of heat energy | Zakarpatgazzbut | 9.5 c.u./$m^3$ |
| Electricity | PJSC Zakarpattyaenergozbut | 3.25 c.u./kWh |
| Cold water supply/ drainage | KP Vodokanal | 37 c.u./$m^3$ |

**Table 2.**
*Information on services provided.*

and determination of the thermal characteristics of its materials, analysis of the heat supply system, assessment of the ventilation system's performance, and evaluation of the hot water supply infrastructure that is presented in **Tables 1** and **2**. These tasks were implemented with strict reference to applicable Ukrainian regulatory documents and methodological guidelines for climate zone II, where Uzhhorod is located.

The audit adhered to the following technical and hygienic requirements: the internal air temperature was set between 18 and 20°C, depending on room function, with an average reference temperature of 20°C for the entire building. The minimum acceptable values for the heat transfer resistance of building envelopes, appropriate for residential and public buildings in this climatic zone, were applied. Furthermore, compliance with the required air exchange rate in rooms according to their function, as well as permissible differences between indoor air temperature and the internal surface temperature of enclosing structures, were monitored to ensure sanitary comfort. The specific energy demand limits for educational institutions, also defined for this climatic region, were considered. Materials used for insulation were assessed based on their compliance with thermal performance and fire safety regulations.

The methodology of the audit combined visual assessment with instrumental diagnostics. The physical condition and level of thermal insulation of walls, windows, and doors were carefully evaluated. Air permeability testing was conducted to determine the extent of air infiltration, particularly around external openings. This test was performed selectively, taking into account the characteristics and operational mode of the building. Air leakage was detected using a thermal imager and anemometers. Measurements of relative humidity and $CO_2$ concentration were performed in multiple rooms, alongside the assessment of surface moisture in the building's wall structures. The geometric parameters of the building envelope were established using available technical documentation, working drawings, and laser-based measurements.

All measurements and observations were carried out according to a pre-approved programme of work, using calibrated instruments. Among the key tools employed during the inspection were: a Testo 875-1i thermal imager for identifying thermal irregularities; Testo and PCE anemometers for airflow evaluation; Retrotec air gate and DM-32 manometer for air leakage testing; sensors for $CO_2$ concentration and surface coatings; and a Bosch GLM-80 laser rangefinder for measuring structural dimensions. Data on indoor air temperature and humidity were collected using digital loggers and thermal detectors, and all findings were documented photographically.

The collected data provided a reliable basis for evaluating the energy performance of the facility is shown in the **Table 3**. It also supported the identification of priority measures for thermal modernisation and optimisation of energy consumption, forming a foundation for further project development aimed at preserving the building's cultural value while improving its energy efficiency.

| Data on heating seasons (beginning and end) | | | | | |
|---|---|---|---|---|---|
| **Start** | **End** | **Number of days** | **Average temperature** | **Degree*day** | **Deviation from the standard, %** |
| 15.10.2020 | 15.04.2021 | 183 | 6,0 | 2562 | 11,1 |
| 15.10.2021 | 15.04.2022 | 183 | 6,2 | 2525 | 12 |
| 15.10.2023 | 15.04.2024 | 183 | 7,2 | 2342 | 18,7 |

**Table 3.**
*Information on the state of heating of the building.*

### *3.2.1 Determination of thermal and geometric characteristics of building envelope materials of the building thermal circuit. Visual inspection of the building envelope*

Determination of materials, thermal and geometric characteristics of building envelope materials of the building thermal circuit is presented in **Table 4**.

The results of the inspection revealed the following:

- There are cracks, delamination, wetting and damage to the finishing layer, as well as places of wetting of structures and crumbling of brickwork joints. Some

| Thermal characteristics of building envelopes | | | | |
|---|---|---|---|---|
| **Construction** | **Material** | **Existing insulation** | **The reduced heat transfer resistance, m$^{2}$*K/W** | |
| | | | Fact | Norma |
| External walls (Type 1) | Compound mortar δ = 0.02 m; ceramic brickwork δ = 0.68 m; cement-sand mortar δ = 0.03 m | — | 1,08 | ≥ 2,8 |
| External walls (Type 2) | Compound mortar δ = 0.02 m; ceramic brickwork δ = 0.42 m | — | 0,67 | — |
| Walls in the ground (Type 3) | Compound mortar δ = 0.02 m; reinforced concrete blocks δ = 0.7 m; bitumen δ = 0.01 m | — | 0,84 | — |
| Cold attic | Lime-sand mortar δ = 0.01 m; Clay shingle δ = 0.02 m, Clay shingle δ = 0.02 m, wooden board δ = 0.04 m, closed air gap, wooden board = 0.04 m | Clay δ = 0,1 м | 0,992 | ≥ 4,5 |
| The floor of a heated basement | Ceramic tiles on cement-sand mortar δ = 0.02 m; cement-sand screed δ = 0.05 m; concrete on crushed natural stone δ = 0.1 m; sand for construction work δ = 0.1 m | — | 3,27 | — |
| External windows | Wooden, double-glazed in a split weave | — | 0,38 | ≥ 0,6 |
| External doors | Wooden | — | 0,23 | ≥ 0,5 |

**Table 4.**
*Results of visual inspection of the building envelope.*

areas of the foundation are leaking moisture, which leads to rapid wear and tear of the building envelope. The heat transfer resistance of the external wall structures does not meet the requirements of regulatory documents.

- The window structures in the building have wooden frames with double glazing. The visual inspection revealed inconsistencies in the technology of processing window slopes and damage to the finishing layer, which led to increased air infiltration and damage to the walls. Some windows with damaged double-glazed windows were also observed.

- The ceiling of the unheated attic is in satisfactory condition, but there are areas where construction debris accumulates and minor wetting occurs. During the inspection of the hipped roof, damage to the wind barrier was found, which violates the heat transfer resistance in accordance with the established standards.

The door structures in the building have wooden frames with single glazing. The heat transfer resistance of these doors also does not meet the requirements of the regulations.

A measuring instrument is used to measure the surface temperature, ambient temperature and relative humidity of the air without contact. It also estimates the dew point and the risk of mould formation based on the measured parameters, including surface temperature and ambient temperature.

After that, the structures are inspected with a thermal detector.

The measuring tool is designed to measure surface temperature, ambient temperature and relative humidity without contact. It also determines the condensation temperature and detects the presence of thermal bridges and the risk of mould. In the mould warning mode, the ambient temperature and relative humidity are measured. The dew point is calculated from this data. The surface temperature is also measured. The comparison of the dew point with the surface temperature allows you to assess the risk of mould growth (**Tables 5** and **6**).

The essence of the test is to pass a flow of air through the object to be tested, and after reaching a steady-state flow, the air flow rate and pressure drop between the internal environment of the object and the external environment or between opposite surfaces of the structure are measured. The test is carried out at a lower or higher pressure than outdoor conditions. The measurement results are used to calculate the generalised air permeability characteristics of the object under test.

When checking the air permeability of the building envelope, the places of infiltration of external air are recorded using thermal imaging. The SMN50 indicator measures the

| | |
|---|---|
| Medium temperature | 20,8 |
| Surface temperature | 17,1 |
| Temperature difference | |
| fact: | 3,7 |
| normal: | up 4,0 |
| Risk of mould growth | low |

**Table 5.**
*Data from the surface temperature survey and mould risk assessment in zone 1.*

| | |
|---|---|
| Medium temperature | 20,8 |
| Surface temperature | 14,8 |
| Temperature difference | |
| fact: | 6,0 |
| normal: | up 4,0 |
| Risk of mould growth | mid-range |

**Table 6.**
*Data from the surface temperature survey and mould risk assessment in zone 2.*

number of cubic metres of air that passes through an object within an hour at a pressure difference between the building and the external environment of 50 Pascals. This indicator, obtained during the test, indicates the aerodynamic properties of the building. A pressure difference of 50 Pa simulates the effect of a 9.2 m/s wind on a building.

The infiltration of outside air through the gaps in the joints of window and wall structures is observed. The infiltration of outside air through the sashes of window structures is also observed.

### *3.2.2 Study of the indoor microclimate*

Indoor climate conditions are a key factor influencing the energy performance of buildings, particularly in the context of energy audits and modernisation planning. Parameters such as air temperature, humidity, ventilation rates, and $CO_2$ concentration affect not only the energy consumption of heating and ventilation systems but also directly impact occupant comfort and well-being. Therefore, understanding and analysing indoor microclimate indicators is essential for determining the building's actual thermal behaviour and its compliance with hygienic standards.

Humidity levels were assessed in both the air and on interior wall surfaces, as this parameter significantly influences indoor air quality. According to established norms, optimal relative humidity lies within the range of 40–60%. Deviations from this range may lead to health risks such as the growth of mould and bacteria at high humidity, or dryness of mucous membranes and eye irritation at low levels. The measurements obtained across the analysed spaces show that air humidity fluctuates from 43.3% in the hallway to 54.3% in the studio, indicating that all tested zones remained within the acceptable range (**Table** 7).

| Room | Relative air humidity, % | | Indoor temperature,°C | | Content $CO_2$ |
|---|---|---|---|---|---|
| | fact* | standard | fact* | standard | standard to 500 ppm |
| Room №1 | 47,3 | 40–60 | 17,0 | 20 | 557 |
| Hallway | 43,3 | | 16,1 | | 476 |
| Room№2 | 51,6 | | 16,1 | | 661 |
| Cellar | 47,6 | | 16,2 | | 617 |
| Studio | 54,3 | | 16,8 | | 733 |

**The measured ambient $CO_2$ level is 450 ppm.*

**Table 7.**
*Information on the indoor microclimate.*

Temperature is considered the most representative indicator of thermal comfort and is commonly measured at a height of 1.5 metres from the floor level. The indoor temperature in the audited areas ranged from 16.1°C to 17.0°C, which is slightly below the recommended standard of 20°C for buildings in climate zone II. This temperature drop is particularly evident during nighttime due to the lack of automatic regulation of the heating system. A moderate rise in daytime temperature is influenced by solar gains but not sufficient to meet normative comfort thresholds.

Carbon dioxide ($CO_2$) concentration was used as a proxy for air quality and ventilation effectiveness. The threshold value for acceptable indoor $CO_2$ levels is 500 ppm. In the investigated rooms, $CO_2$ concentrations ranged from 476 ppm in the hallway to 733 ppm in the studio. Notably, Room №1 (557 ppm), Room №2 (661 ppm), the cellar (617 ppm), and the studio (733 ppm) all exceeded the recommended limit, suggesting insufficient ventilation or high occupancy without air exchange. Given that the background ambient $CO_2$ level was 450 ppm, the excess accumulation indoors reflects an ongoing ventilation challenge in the building.

To interpret the air quality data, a classification system based on $CO_2$ levels was applied, ranging from Category I (excellent, ≤350 ppm) to Category IV (very poor, >1100 ppm), which is depicted in **Table 8**. According to this framework, most of the analysed rooms fall into Category III (moderate), while the studio approaches the upper threshold of acceptable values, nearing the "poor" category.

To better visualise the distribution and compliance of $CO_2$ concentration levels across the analysed rooms, **Figure 10** presents a comparative bar chart highlighting deviations from the standard threshold.

The general pattern observed across the building indicates relatively uniform thermal conditions during the day, with a slight decrease in temperature at night due to the absence of adaptive heating control. While the humidity levels remain within the acceptable range, suboptimal temperature and elevated $CO_2$ concentrations in some zones point to the need for improved heating regulation and enhanced ventilation systems. These findings underscore the importance of integrating environmental monitoring into building modernisation strategies to ensure both energy efficiency and indoor comfort.

### *3.2.3 Investigation of the building's heat supply system and electricity consumption*

The building's heating is provided by a centralised boiler house located in a separate building on the grounds close to the main building. In addition, this boiler room is also responsible for heating two other adjacent buildings that also belong to the institution.

The survey assessed the heating system, the effectiveness of the heat transfer medium control, the presence and quality of insulation of the main pipework and risers that run through unheated rooms, and the condition of the pipework and radiators.

| Category I up to 350 ppm | Category II up to 500 ppm | Category II up to 800 ppm | Category IV - more than 1100 ppm |
|---|---|---|---|
| Good | standart | bad | very bad |
| *$CO_2$ concentration within normal limits* | | | |

**Table 8.**
*Conclusions about the indoor climate.*

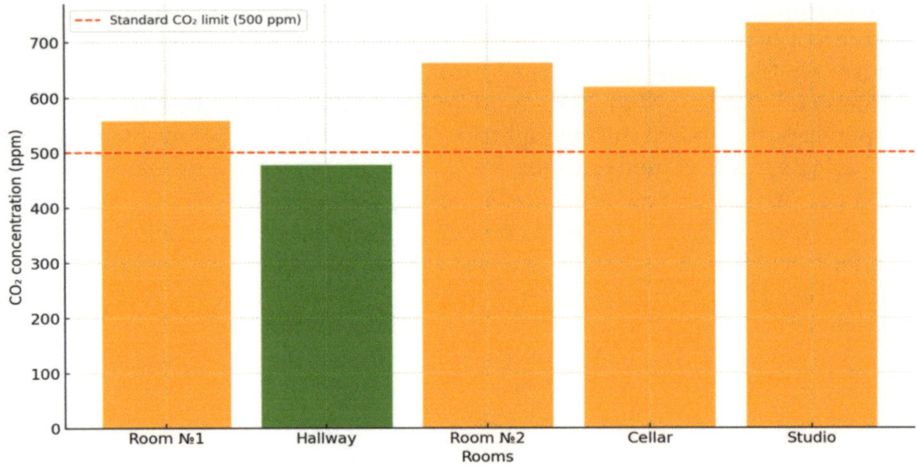


**Figure 10.**
*$CO_2$ concentration levels across the analysed rooms.*

The main pipelines and inlets of the heating system are laid inside the heated basement. The absence of thermal insulation on valves and flanges was detected. The heat transfer devices are contaminated, it is recommended to carry out chemical washing of these devices, and it is also recommended to instal chemical water treatment systems in the boiler room that supplies heat to the building under study.

There is no hot water supply system in this educational institution and no local water heaters. The ventilation of the premises occurs naturally due to the pressure difference between the indoor and outdoor environment, as well as the permeability of the building envelope, such as windows and doors. As for the cooling system, it is also absent, as are local air conditioners.

The lighting system in these premises is a general combined one, i.e. combines natural and artificial lighting, which meets the requirements of the regulations. Natural light is provided by side windows. The condition of the windows in the premises is mostly clean, and there are curtains over the windows. In addition to natural light, artificial electric lighting is also used. The power supply networks for this lighting system are installed concealed, in the thickness of the walls and under the plaster coating, and partially externally in boxes, which meets all the necessary requirements of electrical installation standards. The lighting is controlled in zones, using switches installed on a group of luminaires.

The building's lighting system is divided into interior and exterior lighting. The load distribution of the building's electrical lighting system is shown in **Table 9**.

The main building is supplied with power through the municipal power grid using 0.4 kV cable lines. These lines are laid in pipelines underground from the transformer substation (TS) to the point of entry into the building. The cable coming from the transformer substation is connected to the main input switchboard through the busbars of the input switch. Electricity consumption is metered at the balance sheet responsibility boundary using an electricity metre (**Table 10**).

**Figure 11** compares the total power consumption of various lighting systems in the building before and after a hypothetical replacement with LED technologies. The most significant reductions are observed for luminescent lamps (−72.2%) and incandescent E27 lamps (−68.5%), indicating a high potential for energy savings. In contrast, the LED and outdoor DRL systems remain unchanged. This upgrade

| **Lighting system parameters*** | | | | | |
|---|---|---|---|---|---|
| **Lamp type** | **Lamp power, W** | **Number of lamps per lamp, pcs.** | **The power of the of the luminaire, W** | **The number of of lamps, pcs.** | **Total, W** |
| *Interior lighting* | | | | | |
| LED lamp | 10 | 3 | 30 | 20 | 600 |
| Luminescent Lamp | 36 | 2 | 72 | 213 | 15,336 |
| Incandescent lamps for E27 socket | 60 | 3 | 180 | 35 | 6660 |
| *Outdoor lighting* | | | | | |
| Incandescent lamps for DRL base E42 | 250 | 1 | 250 | 3 | 750 |

**Table 9.**
*Energy audit data for the lighting system.*

| **Significant electricity consumers** | | | | |
|---|---|---|---|---|
| **Name of the equipment** | **Quantity** | **Electrical power of a piece of equipment, kW** | **Total power, kW** | ***Period of use h/day** |
| Muffle furnace | 5 | 4 | 20 | 8 |
| Distillers | 6 | 4,0 | 24 | 6 |
| Fridges | 4 | 0,5 | 2 | 24 |
| Air conditioning | 3 | 1,5 | 4,5 | 8 |
| Printers | 12 | 0,5 | 6 | 8 |
| Computers | 35 | 0,5 | 17,5 | 8 |
| Autoclave | 3 | 6 | 18 | 8 |
| Fume cupboards | 4 | 1 | 4 | 8 |

**Table 10.**
*Information on electricity consumption.*

scenario demonstrates that targeted replacement of inefficient light sources can substantially reduce overall lighting-related electricity consumption in the building.

Based on the data presented in **Table 10**, the total energy consumption of the building's electrical equipment was calculated and summarised in **Table 11**.

The baseline energy consumption of a building is defined as what is expected for the building in accordance with the regulatory requirements for microclimate, internal thermal comfort and building operation conditions (i.e. energy consumption corresponding to the 'baseline condition'). The main parameters of the 'baseline condition' of the microclimate and internal thermal comfort include

- internal air temperature;
- air exchange rate.

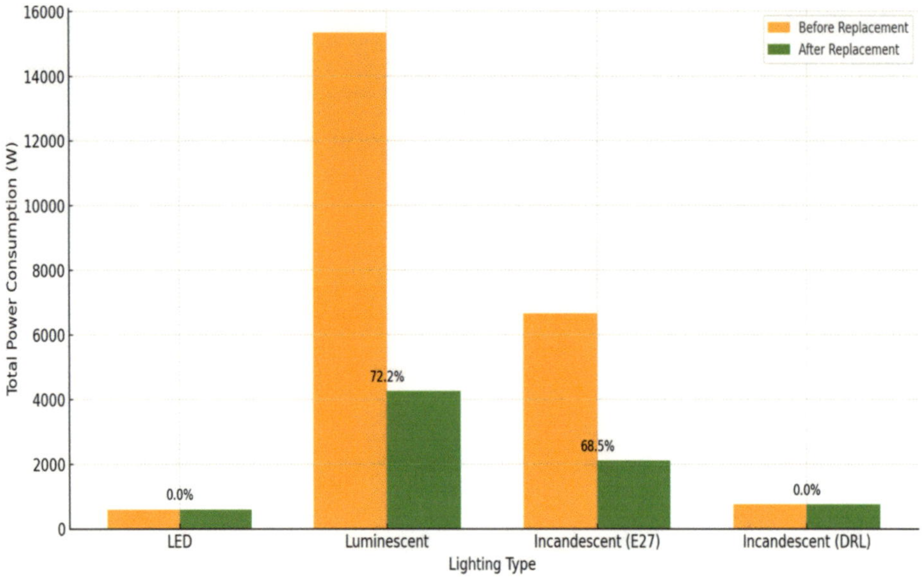


**Figure 11.**
*Comparison of electricity consumption by lighting type before and after replacement.*

| Estimated energy consumption of the building | | | |
|---|---|---|---|
| **Cost item** | **Consumption, kWh** | **Consumption, kWh/year*m$^3$** | **Percentage of of total expenditure** |
| Space heating | 513,329 | 30,5 | 84,5% |
| Ventilation (heating) | | | |
| Hot water supply | — | — | — |
| Lighting | 76,518 | 4,5 | 12,6% |
| Other consumers | 17,352 | 1 | 2,9% |
| Total: | 607,199 | 36 | 100,0% |

**Table 11.**
*Estimated energy consumption.*

Data on baseline energy consumption is shown in **Table 12**.

Based on the data obtained, it is possible to propose measures to modernise the building (**Figure 12**).

The bar chart illustrates the differences between the measured (actual) and expected (baseline) annual electricity consumption for various functional categories in the building, expressed in kWh. The largest savings are observed in lighting (–52.9%) and space heating (–37.4%), indicating successful implementation of energy-saving measures. The complete absence of hot water supply and mechanical ventilation systems also contributes to lower consumption compared to the baseline assumptions. This discrepancy demonstrates the building's reduced operational load, likely resulting from infrastructural limitations or optimisation, and reflects core principles of the circular economy, including reduced energy dependency, extended use of existing infrastructure, and improved efficiency through non-invasive upgrades.

| Baseline energy consumption of the building | | | |
|---|---|---|---|
| **Cost item** | **Consumption, kWh** | **Consumption, kWh/year*m³** | **Percentage of of total expenditure** |
| Heating of premises | 819,711 | 48,6 | 79,6% |
| Ventilation (heating) | | | |
| Hot water supply | 32,130 | 1,9 | 2,7% |
| Lighting | 162,602 | 9,7 | 14% |
| Other consumers | 15,304 | 0,9 | 1,3% |
| Total: | 1,029,747 | 69,3 | 100,0% |

**Table 12.**
*Calculation of the baseline energy consumption of a building.*

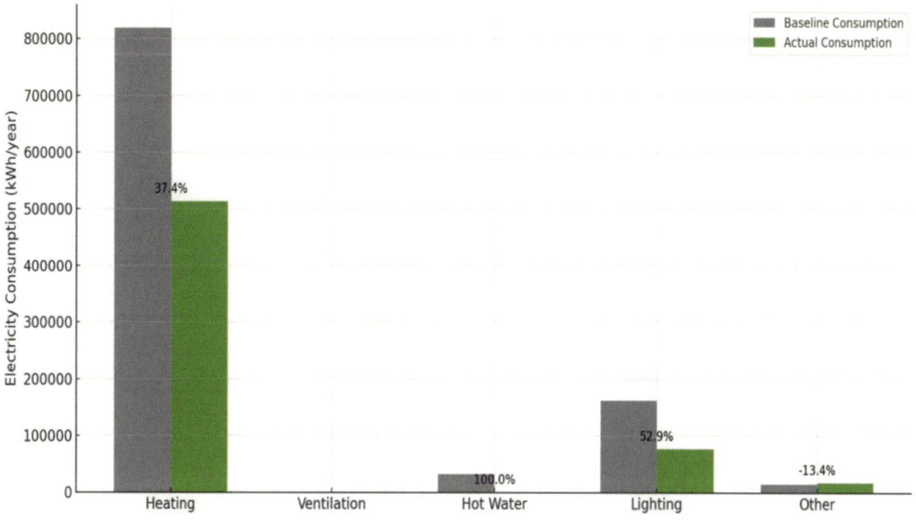


**Figure 12.**
*Actual vs. baseline energy consumption of the building by category.*

## 3.3 Measures to modernise the building

Based on the results of the comprehensive energy audit, a series of modernisation measures were proposed to reduce energy consumption and improve the environmental performance of the analysed cultural heritage building. The selection of these measures was guided by the principles of circular economy — primarily, minimising resource use while extending the lifespan and functionality of existing systems through targeted interventions.

Seven priority actions were identified, addressing both the thermal performance of the building envelope and the operational efficiency of its internal systems. These included insulation of the attic ceiling, replacement of windows and doors, installation of an individual heating point, thermostatic radiator valves, a mechanical ventilation system, and automatic balancing valves on heating risers.

As a result of the analysis of the actual condition of the building and in order to improve the energy efficiency of the building, the following measures and recommendations are proposed below (**Table 13**):

| Indicators of profitability from the implementation of energy efficiency measures | | | | | | | | | |
|---|---|---|---|---|---|---|---|---|---|
| **Name of the event of the event** | **Total savings, kWh/year** | **Cost of implementation, UAH** | **Total savings, UAH/year** | **Pb, years** | **PP, years** | **IRR, %** | **NPV, UAH** | **NPVq** | **The term of operation, years** |
| Installation of an individual heating point | 134,469 | 860,100 | 118,564 | 7,3 | 9,1 | 10,8 | 390,601 | 0,41 | ≤15 |
| Door replacement | 8861 | 78,100 | 8861 | 9,3 | 12.6 | 9,7 | 28,474 | 0,36 | ≤15 |
| Installation of automatic balancing valves | 15,920 | 134,400 | 14,037 | 9,6 | 13,1 | 6,2 | 13,673 | 0,1 | ≤15 |
| Installation of thermostatic regulators | 19,485 | 212,400 | 17,180 | 12,4 | 12,2 | 2,5 | 49,270 | 0,31 | ≤15 |
| Insulation attic ceiling | 120,645 | 1,434,700 | 114,160 | 12,6 | 19,6 | 6,2 | 213,372 | 0,15 | ≤25 |
| Replacement of windows | 105,543 | 2,509,400 | 99,870 | 25,1 | <20 | -5,8 | −1,455,897 | −0,58 | ≤15 |
| Installation of ventilation system | 12,385 | 4,272,800 | 93,010 | 45,9 | <20 | −11,6 | −3,291,661 | −0,77 | ≤15 |
| Total: | 503,220 | 9481,9 | 8,183,400 | 11,6 | — | — | — | — | — |

**Table 13.**
*Preliminary assessment of energy efficiency measures.*

We have assessed the environmental benefits. A preliminary techno-economic analysis was carried out for each proposed measure. **Figure 13** presents the annual energy savings (in kWh/year) expected from each intervention.

Reducing the heat loss of a building leads to a reduction in greenhouse gas emissions. The $CO_2$ emission factor should take into account all emissions associated with all types of energy used. **Table 14** shows the environmental benefits for each of the proposed measures.

Based on the results of the energy audit, a number of energy efficiency measures were proposed and their cost-effectiveness was assessed. The profitability calculations were based on energy tariffs as of March 2020 and took into account a forecast real discount rate of 4.8% (with a nominal discount rate of 10% and an expected inflation rate of 5%). The planning horizon was 20 years. Pb indicates simple payback period; Dp, discounted payback period; n, economic life; NPV, net present value; NPVQ, net present value ratio; IRR, internal rate of return. Conversion factors: 1 kW of heat energy saved = 0.220 kg of $CO_2$.

While the economic analysis demonstrates the profitability of specific interventions, it is equally important to evaluate their environmental impact. **Table 14** provides estimates of annual $CO_2$ emission reductions associated with each measure.

While the previous analysis focused on energy savings and financial performance indicators of modernisation measures, the concept of sustainable renovation must also consider environmental impacts. In line with circular economy principles, reducing the building's energy demand should directly result in a reduction of associated $CO_2$ emissions. **Table 14** presents the estimated environmental benefits of each measure, while **Figure 14** visualises the magnitude of potential greenhouse gas mitigation.

This figure highlights the projected environmental impact of each proposed measure, expressed as the annual reduction of carbon dioxide emissions in tonnes. The largest contributions to emission reduction are expected from the installation of an individual heating point (29.6 t/year), attic insulation (26.5 t/year), and window replacement (23.2 t/year). Together, these three actions represent more than 70% of the total potential $CO_2$ savings. Measures targeting regulation and balancing (e.g. thermostatic valves, balancing valves) offer modest but relevant environmental returns, especially when implemented in combination.

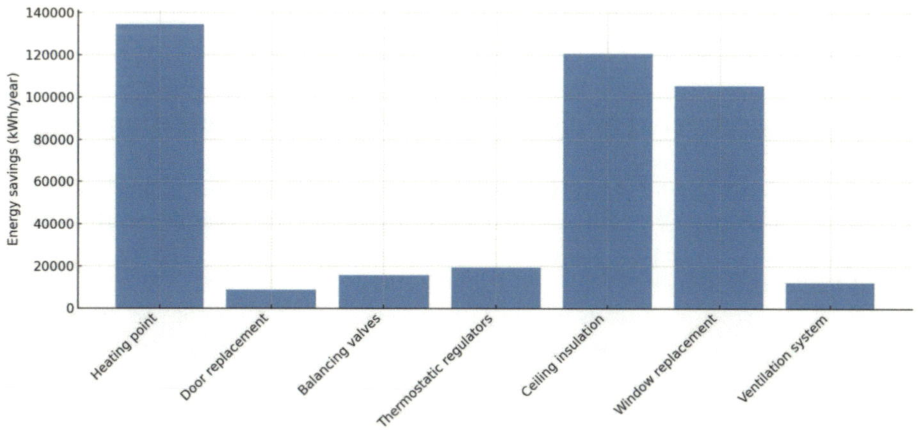


**Figure 13.**
*Annual energy savings from proposed modernisation measures.*

| Environmental benefits from the implementation of measures | | |
|---|---|---|
| **Activities** | **Energy savings, kWh/ year** | **Reduction of $CO_2$ emissions, tonnes per year** |
| Insulation of the attic ceiling | 120,645 | 26,5 |
| Replacement of windows | 105,543 | 23,2 |
| Door replacement | 8861 | 1,95 |
| Installation of an individual heating point | 134,469 | 29,6 |
| Installation of thermostatic regulators on radiators of the heating system | 19,485 | 5,07 |
| Installation of automatic balancing valves | 15,920 | 3,5 |
| Installation of a ventilation system | 12,385 | 21,6 |
| Total: | 503,220 | 111,5 |

**Table 14.**
*Estimated environmental benefits of the measures.*

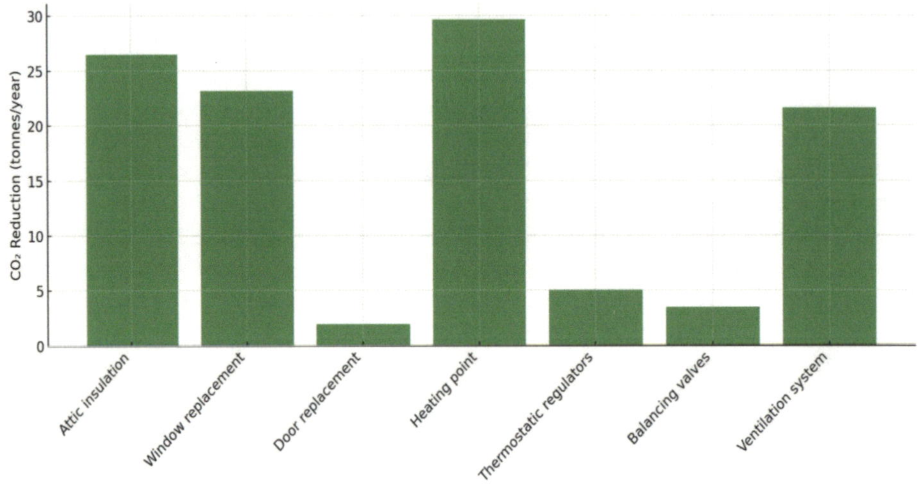


**Figure 14.**
*Estimated annual reduction of $CO_2$ emissions from modernisation measures.*

## 4. Assessing the effect of implementing building modernisation measures

Based on the results of the energy audit (Section 3), a heat energy use balance was compiled. The energy profile of the facility was also created and specific energy efficiency measures were proposed that could reduce the building's energy consumption (**Tables 15** and **16**).

The final stage of the analysis focused on evaluating the effect of the proposed modernisation measures on the building's overall energy efficiency. This assessment was conducted in accordance with the Ukrainian national standard DBN B.2.6–31:2016 Thermal Insulation of Buildings, which stipulates that the annual specific energy demand for educational institutions in climate zone II should not exceed 30 $kWh/m^3$ under standard operational conditions.

| Energy budget of the building | | | | |
|---|---|---|---|---|
| **The item of the budget** | **Estimated to be executed ES** | **Measured (actual values)** | **Prior to the execution of the EA, the base values are** | **After performing the EA** |
| | **[kWh/year]** | **[kWh/ year]** | **[kWh/year]** | **[kWh/year]** |
| *Thermal energy* | | | | |
| Heating | 513,329 | — | 819,711 | 37,902 |
| Ventilation | | | | |
| Hot water supply | — | — | 32,130 | 32,130 |
| *Electric energy* | | | | |
| Lighting | 76,518 | — | 162,602 | 162,602 |
| Other | 17,352 | — | 15,304 | 16,927 |
| **Total** | 607,199 | — | 1,029,747 | 527,150 |

**Table 15.**
*Energy budget of the building.*

| **The item of the budget** | **Estimated to be executed ES** | **Measured (actual values)** | **Prior to the execution of the EA, the base values are** | **After performing the EA** |
|---|---|---|---|---|
| | **[kWh/m$^3$year]** | **[kWh/ m$^3$year]** | **[kWh/m$^3$year]** | **[kWh/m$^3$year]** |
| *Thermal energy* | | | | |
| Heating | 30,5 | — | 48,6 | 18,8 |
| Ventilation | | | | |
| Hot water supply | — | — | 1,9 | 1,9 |
| *Electric energy* | | | | |
| Lighting | 4,5 | — | 9,7 | 9,7 |
| Other | 1 | — | 0,9 | 0,9 |
| **Total** | 36 | — | 69,3 | 31,3 |

**Table 16.**
*Data for calculating the energy budget of a building.*

To assess the impact of modernisation, the energy demand for heating, cooling, and domestic hot water supply was calculated both before and after implementation (**Table 17**). These values were then used to determine the building's energy efficiency class.

**Figure 15** demonstrates a notable reduction in energy demand for space heating, which decreased by more than 50%, from 391,554 kWh/year to 192,770 kWh/year. This confirms the effectiveness of the implemented thermal retrofit measures. The energy consumption for cooling and domestic hot water remained nearly unchanged,

| Criterion | Energy consumption kWh/year | |
|---|---|---|
| | **Basic** | **After the events** |
| Energy demand for heating | 391,554 | 192,770 |
| Energy demand for cooling | 34,158 | 35,548 |
| Energy demand for domestic hot water | 47,824 | 47,824 |
| **Criterion** | **Specific energy consumption for heating, cooling and hot water** | |
| | **Basic** | **After the events** |
| The calculated value, kWh/m$^2$ | 28,1 | 16,4 |
| Energy efficiency class | C | B |

**Table 17.**
*Calculation of the energy efficiency class of a building.*

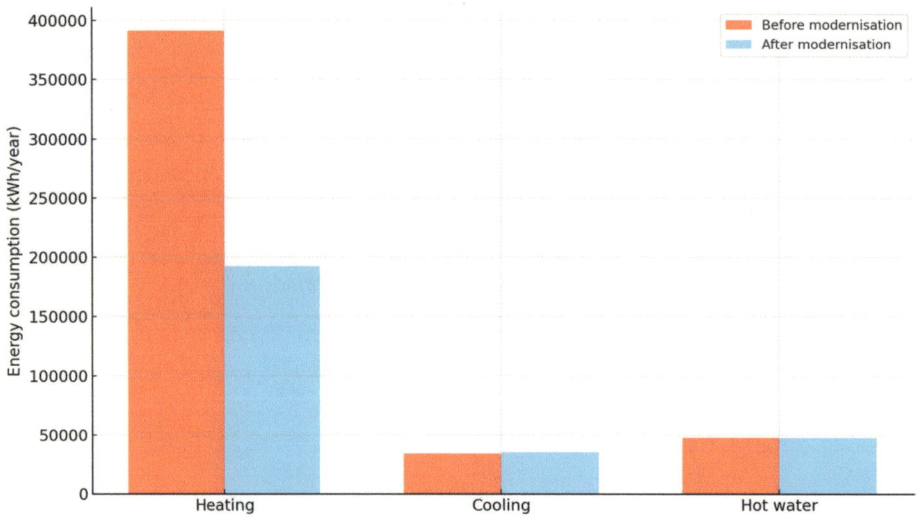


**Figure 15.**
*Energy consumption by type before and after modernisation.*

as these systems were not significantly modified. As a result, the specific energy consumption dropped from 28.1 kWh/m$^2$ to 16.4 kWh/m$^2$, thereby improving the building's energy efficiency class from C to B (**Table 17**). This reclassification signifies enhanced thermal performance and brings the building closer to compliance with the EU Energy Performance of Buildings Directive targets.

Following the implementation of energy efficiency measures, the energy consumption profile of the building was recalculated to reflect the expected operational performance under normative conditions. The recalculations were based on standard thermal comfort parameters and assumed the maintenance of proper microclimatic conditions, including heating, ventilation, and domestic hot water supply. **Figure 16** previously demonstrated significant reductions in space heating demand, which translated into a lower overall specific energy consumption for the building—from 28.1 kWh/m$^2$ down to 16.4 kWh/m$^2$. This improvement resulted in the reclassification of the building's energy efficiency label from Class C to Class B, reflecting compliance

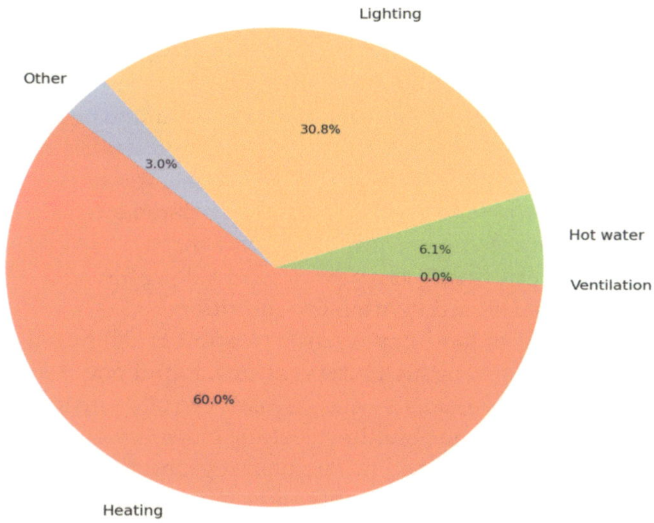


**Figure 16.**
*Structure of energy consumption after modernisation.*

with EU-aligned energy performance criteria for educational buildings in climate zone II (Emax ≤30 kWh/$m^2$). To understand how energy is distributed after the retrofit, **Figure 5** presents the updated structure of energy consumption by use category.

This pie chart highlights the proportional distribution of energy consumption after the implementation of all modernisation measures. Heating remains the dominant category, accounting for approximately 60% of total energy use, followed by lighting (≈31%), and hot water preparation (≈6%). The contribution of ventilation and other internal equipment is negligible.

Such a structure is indicative of a well-insulated and balanced building, where the remaining loads are largely associated with end-use equipment and comfort lighting. Importantly, the new configuration significantly shifts the building's energy profile away from thermal losses and towards functional energy consumption.

## 5. Discussion and conclusions

This study presents a comprehensive assessment of the energy modernisation potential of a historic educational building—the main facility of the Educational and Research Institute of Chemistry and Ecology of Uzhhorod National University. The building, classified as a cultural heritage site of local importance, was selected as a model object for demonstrating how targeted retrofitting measures can reconcile architectural preservation with contemporary requirements for energy performance and environmental sustainability.

The implemented strategy was based on a combination of non-invasive thermal insulation, system-level modernisation (e.g. individual heating point, balancing valves, and thermostatic regulators), and partial envelope upgrades. The evaluation covered three core dimensions: technical performance, economic viability, and environmental impact—aligned with the key principles of circular economy. The results indicate a significant improvement in energy performance. Annual energy consumption was reduced by over 41%, with a shift in the building's energy efficiency class from C to B, as per DBN B.2.6-31:2016 standards. The calculated specific energy consumption decreased from 28.1 to 16.4 kWh/m$^2$, placing the building well within national regulatory thresholds. Furthermore, the projected annual reduction in $CO_2$ emissions exceeds 111 tonnes, contributing to regional climate neutrality goals.

In addition to environmental benefits, the study demonstrated clear economic advantages. Several measures showed attractive internal rates of return (e.g. IRR > 10%) and short payback periods, while others offered long-term strategic value, especially when bundled into comprehensive renovation packages. The structure of energy use after retrofitting revealed a balanced profile, with reduced reliance on heating systems and optimised lighting consumption.

The contextual analysis of landscape transformation in Uzhhorod (2017–2024) further underlines the urgency of such interventions. Rapid population growth—exacerbated by wartime displacement—has stressed the city's infrastructure, particularly the residential and public building sectors. However, cultural heritage sites remain largely overlooked in state and municipal programmes. The demonstrated modernisation framework offers a replicable and scalable model for integrating such buildings into broader urban resilience and sustainability policies. In summary, this study confirms that energy retrofitting of historic buildings is not only technically and economically feasible but can be strategically leveraged as part of national decarbonisation efforts. The methodology and results presented here may serve as a foundation for the development of standardised renovation protocols for heritage buildings in Eastern Europe and beyond, contributing to both the preservation of cultural identity and the achievement of climate goals.

## 6. Limitations of the study and future research directions

While this study provides an in-depth assessment of energy efficiency measures in a representative cultural heritage building, several limitations must be acknowledged. First, the analysis was conducted on a single pilot site—the Educational and Research Institute of Chemistry and Ecology of Uzhhorod National University. Although this case study offers valuable insights, its generalisability remains limited due to the lack of a systematic state-supported database or registry of cultural heritage buildings in Ukraine.

The absence of a dedicated national framework for the energy modernisation of heritage assets constrains the scalability of such interventions. Currently, the primary financial drivers behind these projects are international donors and decentralised municipal efforts. The ongoing military conflict further complicates the regulatory and infrastructural environment, limiting institutional capacity, disrupting supply chains, and shifting national priorities away from heritage conservation and energy transition efforts.

From a methodological perspective, this study focused on energy modelling, economic indicators, and qualitative assessments of indoor environmental quality. However, more granular and longitudinal data are needed to fully evaluate the comfort, health, and resilience benefits of retrofitting. In particular, continuous monitoring of indoor air quality, thermal comfort, and real-time energy use is essential to validate modelling assumptions and understand occupant behaviour in heritage buildings. Future research should therefore focus on:

1. Expanding the dataset to include a statistically significant sample of buildings with varying typologies, historical classifications, and climate exposures;

*DOI: http://dx.doi.org/10.5772/intechopen.1010319*

2. Implementing life cycle cost assessments (LCCA) and life cycle carbon footprint analysis (LCA) to quantify long-term sustainability impacts of retrofit scenarios;

3. Investigating material compatibility and heritage-sensitive design solutions, particularly for insulation, window replacement, and renewable energy integration;

4. Developing a framework for hybrid performance evaluation, combining quantitative energy and environmental metrics with qualitative assessments of cultural value retention;

5. Informing and co-developing with policymakers a national roadmap for sustainable modernisation of cultural heritage buildings, with Uzhhorod serving as a potential pilot city and policy laboratory.

Ultimately, the goal of future research is to offer a scientifically grounded and culturally sensitive methodology for improving the energy performance of historic buildings, while preserving their architectural integrity and enhancing their long-term contribution to climate neutrality strategies.

## Acknowledgements

The authors wish to express sincere gratitude to Mr R.Z. Saliuk, Director of LLC "Adamson Service Company," for his organisational support and practical assistance in the implementation of this research project.

We are also grateful to the energy efficiency experts Mr Ye.H. Bryzhytskyi, Mr D.V. Litvinov, and Mr M.V. Stashko for their professional consultations, technical expertise, and valuable contributions to the assessment of energy-saving measures applied to the cultural heritage site.

## Conflict of interest

The authors declare no conflict of interest.

## Author details

Sérgio Lousada[1,2,3,4,5*], Andrii Khorolskyi[6], Myroslava Lendel[7], Oksana Sviezhentseva[8] and Svitlana Delehan[9]

1 Department of Civil Engineering and Geology (DECG), Faculty of Exact Sciences and Engineering (FCEE), University of Madeira (UMa), Funchal, Portugal

2 VALORIZA – Research Centre for Endogenous Resource Valorization, Portalegre, Portugal

3 RISCO – Civil Engineering Department of University of Aveiro, Aveiro, Portugal

4 CITUR – Madeira – Centre for Tourism Research, Development and Innovation, Madeira, Portugal

5 Research Group on Environment and Spatial Planning (MAOT), University of Extremadura, Badajoz, Spain

6 Laboratory of Mining Problems, Branch for Physics of Mining Processes of the National Academy of Sciences of Ukraine, Dnipro, Ukraine

7 Department of Political Science and Public Administration, Uzhhorod National University, Uzhhorod, Ukraine

8 Department of International Studies and Public Communications, Uzhhorod National University, Uzhhorod, Ukraine

9 Centre for Interdisciplinary Scientific Research, Uzhhorod National University, Uzhhorod, Ukraine

*Address all correspondence to: slousada@staff.uma.pt

## References

[1] Naheed S, Shooshtarian S. The role of cultural heritage in promoting urban sustainability: A brief review. Land. 2022;**11**(9):1508. DOI: 10.3390/land11091508

[2] Buildings Performance Institute Europe. Europe's Buildings under the Microscope. A Country-By-Country Review of the Energy Performance of Buildings. Bruxelles: European Commission; 2011

[3] Council of Europe. European cultural convention. 1954. Available from: https://www.coe.int/en/web/culture-and-heritage/european-cultural-convention

[4] Birchall S et al. Survey on the energy needs and architectural features of the EU building stock Deliverable 2.1a, European Project iNSPIRe. 2014. Available from: www.inspirefp7.eu

[5] European Parliament. Directive 2010/31/EU of the European parliament and of the council of 19 May 2010 on the energy performance of buildings (recast), Official Journal of the European Union; 2010

[6] European Parliament. Directive 2009/28/EC on the promotion of the use of energy from renewable sources, Official Journal of the European Union; 2009

[7] European Parliament. Directive 2012/27/EU on energy efficiency, Official Journal of the European Union; 2012

[8] European Committee for Standardization (CEN). Indoor Environmental Parameters for Assessment of Energy Performance of Buildings, Addressing Indoor Air Quality, Thermal Environment, Lighting and Acoustics, Standard EN 15251. Brussels, Belgium: CEN; 2007

[9] European Committee for Standardization (CEN). Conservation of Cultural Property. Specifications for Temperature and Relative Humidity to Limit Climate-Induced Mechanical Damage in Organic Hygroscopic Materials, Standard EN 15757. Brussels, Belgium: CEN; 2010

[10] European Committee for Standardization (CEN). Conservation of Cultural Property. Procedures and Instruments for Measuring Temperatures of the Air and the Surfaces of Objects, Standard EN 15758. Brussels, Belgium: CEN; 2010

[11] European Committee for Standardization (CEN). Conservation of Cultural Property. Indoor Climate, Standard EN 15759. Brussels, Belgium: CEN; 2011

[12] European Committee for Standardization (CEN). Conservation of Cultural Heritage. Procedures and Instruments for Measuring Humidity in the Air and Moisture Exchanges between Air and Cultural Property, Standard EN 16242. Brussels, Belgium: CEN; 2012

[13] European Committee for Standardization (CEN). Conservation of Cultural Heritage. Guidelines and Procedures for Choosing Appropriate Lighting for Indoor Exhibitions, Standard CEN/TS 16163. Brussels, Belgium: CEN; 2014

[14] Lucchi E. Energy efficiency of historic buildings. 2018. Available from: https://www.semanticscholar.org/paper/Energy-Efficiency-of-Historic-Buildings-

Lucchi/6963ffd6ad7afbd8262b228d7e2032 06a6975d25

[15] Bullen P. Adaptive reuse of heritage buildings: Sustaining an icon or eyesore. In: Proceedings of the COBRA 2011—Proceedings of RICS Construction and Property Conference. Salford, UK; 12-13 September 2011. pp. 1652-1662. (PDF) Establishment of Austro-Hungarian Military Buildings in Banja Luka and Their Subsequent Repurposing. Available from: https://www.researchgate.net/publication/386404030_Establishment_of_Austro-Hungarian_Military_Buildings_in_Banja_Luka_and_Their_Subsequent_Repurposing [Accessed: April 29, 2025]

[16] Pérez-Lombard L, Ortiz J, Pout C. A review on buildings energy consumption information. Energy and Buildings. 2008;**40**(3):394-398

[17] Pyshchulina O, Yurchyshyn V, Stetsiuk P, Mishchenko M. Vplyv mihratsiinykh protsesiv na perebih povojennoho vidnovlennia Ukrainy, mihratsiinu polityku YeS ta krain Skhidnoho Partnerstva. In: Yakymenko Y, editor. Kyiv: Razumkov Centre; 2023. 265 p

[18] Garzulino A. Energy Efficiency: A multi-criteria evaluation method for the intervention on built heritage. Sustainability. 2020;**12**(21):9223. DOI: 10.3390/su12219223

[19] Carfora A, Scandurra G. Boosting green energy transition to tackle energy poverty in Europe. Energy Research and Social Science. 2024;**103**:103451. DOI: 10.1016/j.erss.2024.103451

[20] Analytics International Security. Hague convention cultural property Ukraine. Available from: https://analytics.intsecurity.org/en/hague-convention-cultural-property-ukraine/

[21] Ministry of Culture and Information Policy of Ukraine. Cultural heritage of Ukraine: 70 years of the Hague convention for the protection of cultural property. Available from: https://mcsc.gov.ua/news/kulturna-spadshhyna-ukrayiny-70-richchya-gaazkoyi-konvencziyi-pro-zahyst-kulturnyh-czinnostej-u-vypadku-zbrojnogo-konfliktu/

[22] TSN. Ukraine creates a military unit for the protection of historical heritage. Available from: https://tsn.ua/ato/v-ukrayini-stvoreno-pidrozdil-zsu-dlya-zahistu-istorichnoyi-spadschini-podrobici-2728794.html

[23] ICOMOS Archive. Recommendations for the protection of cultural heritage in Ukraine. Available from: https://openarchive.icomos.org/id/eprint/3356/1/%D1%80%D0%B5%D0%BA%D0%BE%D0%BC%D0%B5%D0%BD%D0%B4%D0%B0%D1%86%D1%96%D1%97%D0%B3%D1%80%D1%83%D0%BF%D0%B8%20%D0%B5%D0%BA%D1%81%D0%BF%D0%B5%D1%80%D1%82%D1%96%D0%B2%20%D0%BF%D1%96%D0%B4%20%D0%BA%D0%B5%D1%80%D1%96%D0%B2%D0%BD%D0%B8%D1%86%D1%82%D0%B2%D0%BE%D0%BC%20%D1%94%D0%B2%D1%80%D0%BE%D0%BF%D0%B5%D0%B9%D1%81%D1%8C%D0%BA%D0%BE%D1%97-NC0224704UKN.pdf

[24] Padfield T. The role of standards and guidelines in cultural heritage protection. In: Krumbein WE, editor. Durability and Change. Copenhagen, Denmark: Wiley; 1994. pp. 191-199

[25] CEN/TC 346. Conservation of cultural heritage. European Committee for Standardization: Brussels, Belgium. Available from: https://standards.iteh.ai/catalog/tc/cen/782ad083-d5d4-4d4f-ac6d-36572d262c15/cen-tc-346

[26] Quattrone G. Revitalization practices for resilient cities. Creative potential of heritage regarding revitalization and renewal of cities. Procedia Environmental Sciences. 2017;**37**:466-473

[27] Svitlana D, Hanna M. Environmental and energy security on the way to rebuilding Ukraine on the principle of zero carbon footprint. In: Nate S, editor. Ukraine's Journey to Recovery, Reform and Post-War Reconstruction. Cham: Springer; 2025. pp. 203-215. DOI: 10.1007/978-3-031-66434-2_14

[28] Verkhovna Rada of Ukraine. Law of Ukraine "On the Protection of Cultural Heritage", No. 1805-III, adopted on 08.06.2000. Available from: https://zakon.rada.gov.ua/laws/show/1805-14 [Accessed: 20 March 2025]

[29] Verkhovna Rada of Ukraine. Law of Ukraine "On Architectural Activity", No. 687-XIV, adopted on 20.05.1999. Available at: https://zakon.rada.gov.ua/laws/show/687-14 [Accessed: 20 March 2025]

[30] Verkhovna Rada of Ukraine. Law of Ukraine "On Investment Activity", No. 1560-XII, adopted on 18.09.1991. Available from: https://zakon.rada.gov.ua/laws/show/1560-12 [Accessed: 20 March 2025]

[31] Verkhovna Rada of Ukraine. Law of Ukraine "On Regulation of Urban Development Activities", No. 3038-VI, adopted on 17.02.2011. Available from: https://zakon.rada.gov.ua/laws/show/3038-17 [Accessed: 20 March 2025]

[32] Delehan S. Applying a Zero Carbon Footprint Strategy to Improve Environmental and Energy Security. London, UK: IntechOpen; 2025

[33] Svitlana D, Malychkovych O, Hanna M. Study of the state of environmental information security of Ukraine in the pre-war period and assessment of the consequences of environmental information manipulations in view of changes in the geopolitical landscape and new strategic risks. In: Nate S, editor. Ukraine's Journey to Recovery, Reform and Post-War Reconstruction. Cham: Springer; 2025. pp. 217-228. DOI: 10.1007/978-3-031-66434-2_15

[34] Loures L, Castanho RA, Gómez JMN, Lousada SA, Fernández-Pozo L, Cabezas J, et al. Impactos Socioculturais da Cooperação Transfronteiriça (CT) no Espaço Europeu. Fronteiras. 2019;**8**(3):292-312. DOI: 10.21664/2238-8869.2019v8i3.p. 292-312

[35] Lousada S, Delehan S, Naranjo Gómez JM, Gallardo JM, Mandryk O, Khorolskyi A. Bridging the gap: Overcoming the 85 mm railway gauge difference between Ukraine and Europe using principles of circular economy and European service quality standards. Journal of Infrastructure, Policy and Development. 2024;**8**(16):10555. DOI: 10.24294/jipd10555

[36] Lousada S, Delehan S, Khorolskyi A. Application of dynamic programming models for improvement of technological approaches to combat negative water leakage in the underground space. Water. 2024;**16**(14):1952. DOI: 10.3390/w16141952

[37] Lousada S, Delehan S, Vilčekova S, Khorolskyi A. Applying a Zero Carbon Footprint Strategy to Improve Environmental and Energy Security. London, UK: IntechOpen; 2025. DOI: 10.5772/intechopen.1007890

[38] Lefebvre A, Sannier C, Corpetti T. Monitoring urban areas with sentinel-2A

data: Application to the update of the Copernicus high resolution layer imperviousness degree. Remote Sensing. 2016;**8**(7):606. DOI: 10.3390/rs8070606

[39] Phiri D, Simwanda M, Salekin S, Nyirenda VR, Murayama Y, Ranagalage M. Sentinel-2 data for land cover/use mapping: A review. Remote Sensing. 2020;**12**(14):2291. DOI: 10.3390/rs12142291

[40] Kranjčić N, Medak D, Župan R, Rezo MJ. Support vector machine accuracy assessment for extracting green urban areas in towns. Remote Sensing. 2019;**11**:655. DOI: 10.3390/rs11060655

[41] Hossain ML, Li J. NDVI-based vegetation dynamics and its resistance and resilience to different intensities of climatic events. Global Ecology and Conservation. 2021;**30**:e01768. DOI: 10.1016/j.gecco.2021.e01768

[42] Barbosa R, Vicente R, Santos R. Climate change and thermal comfort in southern Europe housing: A case study from Lisbon. Building and Environment. 2015;**92**:440-451. DOI: 10.1016/j.buildenv.2015.05.019

[43] Khorolskyi A, Hrinov V, Mamaikin O, Fomychova L. Research into optimization model for balancing the technological flows at mining enterprises. E3S Web of Conferences. 2020;**201**:01030. DOI: 10.1051/e3sconf/202020101030

[44] Lousada S, Cabezas J, Castanho RA, Gómez JMN. Land-use changes in insular urban territories: A retrospective analysis from 1990 to 2018. The case of Madeira Island—Ribeira Brava. Sustainability. 2022;**14**(24):16839. DOI: 10.3390/su142416839

[45] Lousada S, Silva PSD, Castanho RA, Naranjo-Gómez JM. Modelação de sistemas de abastecimento de água. O caso de Ilha da Madeira. Bitácora Urbano Territorial. 2019;**29**(2):89-98. DOI: 10.15446/bitacora.v29n2.69381

[46] Delehan S, Melehanych H, Khorolskyi A. The traditions and technologies of ecological construction in Portugal. Engineering Proceedings. 2023;**57**(1):23. DOI: 10.3390/engproc2023057023

[47] Van Den Bergh JCJM, Grazi F. Reply to the first systematic response by the global footprint network to criticism: A real debate finally? Ecological Indicators. 2015;**58**:458-463. DOI: 10.1016/j.ecolind.2015.05.007

[48] Kaposztasova D, Lavkova Cakyova K, Vertal M, Vranayova Z, Burdova EK. Active green constructions and their impact on gray infrastructure. Buildings. 2024;**14**(2):306. DOI: 10.3390/buildings14020306

[49] Ministry of Regional Development, Construction and Housing and Communal Services of Ukraine. DBN B.2.6-31:2016. Thermal Insulation of Buildings. Kyiv: Minregion; 2016

[50] Ministry of Regional Development and Construction of Ukraine. DSTU-N B V.1.1-27:2010. Guidelines for the Determination of the Composition of Thermal Protection Measures of Buildings. Kyiv: Minregion; 2010. (In Ukrainian)

[51] Ministry of Regional Development of Ukraine. DSTU B B.2.2-39:2016. Methods and Stages of Energy Audit of Buildings. Kyiv: Minregion; 2016

[52] Ministry of Regional Development of Ukraine. DSTU B B.2.2-19:2007. Method for Determining the Air Permeability of Building Envelopes

in Natural Conditions. Kyiv: Minregion; 2007

[53] Ministry of Regional Development of Ukraine. DSTU-N B B.2.6-191:2016. Guidelines for the Calculation of Air Permeability of Building Envelopes. Kyiv: Minregion; 2016

[54] Ministry of Regional Development of Ukraine. DSTU B EN 13187:2011. Thermal Performance of Buildings. Qualitative Detection of Thermal Failures in Building Envelopes. Infrared Method. Kyiv: Minregion; 2011

[55] Ministry of Regional Development of Ukraine. DSTU B EN 15251:2011. Estimated Parameters of the Indoor Microclimate for the Design and Assessment of the Energy Performance of Buildings in Terms of Air Quality, Thermal Comfort, Lighting and Acoustics. Kyiv: Minregion; 2011

[56] Ministry of Regional Development of Ukraine. DSTU B a.2.2-12:2015. Energy Efficiency of Buildings. Method for Calculating Energy Consumption for Heating, Cooling, Ventilation, Lighting and Hot Water Supply. Kyiv: Minregion; 2015

[57] Ministry of Regional Development of Ukraine. DSTU B EN 15603:2013. Energy Efficiency of Buildings. Total Energy Consumption and Energy Assessment (EN 15603:2008, IDT). Kyiv: Minregion; 2013

[58] Ministry of Regional Development of Ukraine. DSTU B EN 15217:2012. Energy Efficiency of Buildings. Methods of Presenting Energy Performance and Energy Certification of Buildings. Kyiv: Minregion; 2012

[59] Ministry of Regional Development of Ukraine. DSTU B EN 15316-1:2011. Heat Supply Systems for Buildings. Methodology for Calculating Energy Consumption and Energy Efficiency of the System. Kyiv: Minregion; 2011

[60] Ministry of Regional Development of Ukraine. DBN B.2.5-64:2012. Internal Water Supply and Sewerage. Kyiv: Minregion; 2012

[61] Ministry of Regional Development of Ukraine. DBN B.2.5-67:2013. Heating, Ventilation and Air Conditioning. Kyiv: Minregion; 2013

[62] Ministry of Economic Development of Ukraine. DSTU ISO 50002:2016. Energy Audits. Requirements and Guidelines for their Conduct. Kyiv: MEDT; 2016

[63] Ministry of Economic Development of Ukraine. DSTU ISO 50004:2016 (ISO 50004:2014, IDT). Guidelines for the Implementation, Maintenance and Improvement of the Energy Management System. Kyiv: MEDT; 2016

[64] Ministry of Regional Development of Ukraine. DSTU B B.2.6-189:2013. Methods of Selecting Heat-Insulating Material for Building Insulation. Kyiv: Minregion; 2013

[65] Ministry of Regional Development of Ukraine. DSTU-N B V.2.6-146:2010. Constructions of Buildings and Structures. Guidelines for the Design and Installation of Windows and Doors. Kyiv: Minregion; 2010

[66] Ministry of Regional Development of Ukraine. DSTU-N B a.2.2-13:2015. Guidelines for Energy Assessment of Buildings. Kyiv: Minregion; 2015

[67] Ministry of Regional Development of Ukraine. DSTU B EN ISO 13790:2011. Energy Efficiency of Buildings. Calculation of Energy Consumption

for Heating and Cooling (EN ISO 13790:2008, IDT). Kyiv: Minregion; 2011

[68] Ministry of Regional Development of Ukraine. DBN B.2.5-28:2006. Engineering Equipment of Buildings and Structures. Natural and Artificial Lighting. Kyiv: Minregion; 2006

[69] Ministry of Regional Development of Ukraine. DSTU B B.2.6-101:2010. Structures of Buildings and Structures. Method for Determining the Heat Transfer Resistance of Building Envelopes. Kyiv: Minregion; 2010

Chapter 9

# Urban Design and Planning: Integrating Nature and the Built Environment

*Deborah Bunmi Ojo*


## Abstract

This chapter will investigate the critical function of urban design and planning within the larger context of landscape architecture and design. The chapter will look at the concepts, procedures, and practices that drive successful urban design and planning initiatives from a variety of perspectives. The chapter will also explore the changing difficulties and opportunities confronting contemporary urban environments, such as growing urbanization, climate change, social inequality, and technological breakthroughs. It emphasizes the importance of using innovative and adaptive urban planning and design strategies that promote equity, resilience, and environmental stewardship. Using a variety of case studies and best practices, the chapter will highlight remarkable urban design and planning initiatives from around the world. These examples will show how purposeful design interventions can greatly improve the standard of living for city residents while also fostering sustainable urban expansion.


**Keywords:** built environment, landscape architecture, planning, sustainability, urban design

## 1. Introduction

Urban design and planning are pivotal disciplines within the broader field of Landscape Architecture and Design, profoundly shaping the environments where we live, work, and socialize. This chapter explores the symbiotic relationship between natural landscapes and urban structures, emphasizing the imperative for cohesive and sustainable development practices. Traditionally, architectural theory has underpinned the evolution of urban design and planning, which operate synergistically within the same physical space [1]. This study examines how integrating these fields enhances the quality and form of cities in practical application. Urban design focuses specifically on the physical form and esthetic quality of urban spaces, encompassing the design and arrangement of buildings, public spaces, streets, and landscapes to create functional, attractive, and cohesive urban environments [2, 3]. Urban designers work across scales from individual projects to neighborhoods and city districts—aiming to enhance user experience and foster a distinct sense of place. On the other hand, urban planning entails a holistic process that entails creating and executing policies, strategies, and plans to regulate land use, infrastructure, and services within urban areas [4, 5]. It spans activities such as zoning, transportation planning, environmental protection, economic development, and community engagement, with the primary goal of ensuring orderly growth and addressing social, economic, and environmental needs to enhance resident quality of life.

While distinct in their focus, the integration of urban planning and design is crucial for creating well-functioning and sustainable urban areas. Urban planners establish frameworks and guidelines that guide urban designers, ensuring alignment with broader goals and policies. Effective collaboration between these disciplines leads to cohesive and harmonious urban environments that meet present and future needs. For instance, urban planners might identify the need for new housing, while urban designers conceptualize how these units integrate into the urban fabric, ensuring they are not only functional and compliant but also esthetically pleasing and supportive of community well-being. The practice of urban design and planning has increasingly embraced the integration of natural elements into urban spaces as a strategy for creating sustainable, livable, and resilient cities. Currently, more than half of the world's population lives in urban areas, which cover more than 3% of the global ecosystem and are expanding rapidly due to construction growth [6]. This unprecedented urbanization poses challenges but also opportunities to harmonize nature with the built environment. Urban environments significantly impact human and environmental health, underscoring the importance of integrating natural systems such as green spaces, urban agriculture, wildlife corridors, and sustainable water and waste management practices [7–9].

This cohesive strategy not only boosts ecological resilience and biodiversity but also aids in enhancing air quality, mitigating urban heat islands, and improving stormwater management. It reflects a shift from historical city designs that separated built-up areas from green spaces towards modern principles that recognize the benefits of natural systems in enhancing urban liveability and sustainability [10, 11]. Urban planners and designers play pivotal roles in advancing these practices, collaborating with diverse stakeholders to resolve challenges and create environments that balance human needs with environmental stewardship. This introduction sets the stage for examining how urban design and planning integrate with natural systems to create sustainable and liveable urban environments, emphasizing their critical roles in shaping the future of cities worldwide.

## 2. Historical perspectives on urban design and planning

The evolution of urban design and planning has been significantly influenced by various movements, theories, and practices over time. From ancient civilizations to contemporary urban challenges, each era has contributed unique perspectives and methodologies that continue to influence how cities are designed and developed. Ancient urban planning principles laid the groundwork for organized city layouts that balanced functionality with societal needs. In Mesopotamia and classical Greece, gridiron street patterns and centralized public spaces reflected a systematic approach to urban organization: these early developments emphasized the importance of order and efficiency in urban living, establishing enduring principles that would evolve over centuries. During the Renaissance and Baroque periods, urban planning became a medium for expressing political power and cultural identity. Cities such as Florence and Rome were transformed with grand boulevards, monumental architecture, and expansive public squares that served as civic symbols. This era emphasized principles of symmetry and hierarchy, integrating esthetics with urban functionality to create environments that inspired awe and civic pride [12–18].

The Enlightenment era marked a shift towards urban planning focused on public health and sanitation in response to rapid industrialization. The introduction of zoning regulations, public parks, and sanitation systems aimed to mitigate the negative effects of urbanization [4]. The Industrial Revolution further reshaped urban landscapes with the proliferation of factories and tenement housing, necessitating new planning strategies to manage urban growth and improve living conditions. The early twentieth century witnessed the rise of modernist movements that sought to redefine urban spaces through functionalist designs and technological innovation. Figures like Le Corbusier envisioned utopian cities characterized by high-rise buildings, green spaces, and efficient transportation networks [19]. These modernist ideals aimed to address urban congestion and social inequality but also sparked critiques regarding their impact on community cohesion and environmental sustainability. Post-World War II urban planning strategies focused on urban renewal and redevelopment to address post-war challenges and stimulate economic growth. Initiatives such as urban renewal projects aimed to modernize infrastructure and alleviate housing shortages but often resulted in the displacement of communities and loss of cultural heritage [20]. This period marked a turning point with increasing calls for more inclusive and participatory planning approaches that prioritize community engagement and social equity.

In the modern era, urban design and planning encounter intricate obstacles including climate change, rapid urbanization, and societal disparities. Concepts such as sustainable development, smart growth, and resilience have arisen as fundamental ideals in crafting cities that are both environmentally sustainable and socially inclusive [21–24]. These contemporary paradigms advocate for compact, walkable communities, green infrastructure, and adaptive reuse strategies that promote long-term urban resilience and liveability.

## 3. Principles of urban design and planning

Urban design principles encompass a range of foundational concepts that guide the planning and development of urban spaces. These principles have evolved over time, influenced by historical precedents, social values, and environmental considerations. These principles highlight the significance of developing cities that are sustainable, resilient, esthetically pleasing, inclusive, and designed to cater to the diverse needs of their residents [25–27]. Here are some key principles:

a. *Human scale*: Designing urban spaces that are comfortable and accessible to people, considering the proportions and dimensions that are suitable for human interaction and activities. This concept, which promotes human scale and communal connection, was inspired by critiques of modernist urban architecture. In her seminal essay "The Death and Life of Great American Cities," Jane Jacobs stressed the importance of diverse public spaces, pedestrian-friendly streets, and mixed-use neighborhoods [20].

b. *Diversity:* Diversity in urban design involves creating mixed-use neighborhoods that integrate residential, commercial, recreational, and institutional components, promoting a vibrant and healthy community life. Encouraging a blend of these uses within neighborhoods fosters dynamic and diverse communities. Additionally, cultural diversity within a city enhances its economic vitality, offering locals a wide array of cultural products and heritage. This multicultural richness also makes the urban landscape more attractive to visitors, boosting tourist [4, 5], and related industries. Such development is crucial for fostering new businesses and economic vibrancy, ultimately leading to increased employment opportunities. Generally, it means maintaining variety and choice

c. *Walkability*: Designing pedestrian-friendly surroundings with connected streets, walkways, and public areas to encourage walking as a key form of transportation. The principle emphasizes the importance of walkable neighborhoods as a way to promote a healthy lifestyle and reduce dependence on automobiles.

d. *Transit-oriented development*: Designing urban areas around public transportation hubs to promote the use of public transit, reduce reliance on cars, and create compact, mixed-use developments.

e. *Green spaces:* The inclusion of parks, gardens, and green infrastructure in urban planning improves air quality, provides recreational opportunities, and enhances residents' overall quality of life. These spaces not only reduce urban heat islands but also enhance biodiversity, support mental well-being, and contribute to essential ecosystem services, fostering sustainable urban environments.

f. *Historic preservation and identity*: Protecting and preserving historic buildings, landmarks, and cultural heritage to maintain a sense of place and continuity with the past.

g. *Community engagement:* Active involvement of residents, stakeholders, and community groups in the planning and design process ensures that urban developments are tailored to meet the needs and aspirations of the local population.

h. *Safety and security*: Designing urban spaces that are safe, well-lit, and easily monitored to reduce crime, enhance public safety, and create a sense of security for residents and visitors.

i. *Adaptability and flexibility*: Designing urban spaces that can adapt to changing needs and circumstances over time, allowing for flexibility in land use, infrastructure, and design elements.

j. *Proximity:* Earlier movements such as the garden city and the neighborhood unit have also emphasized proximity as a fundamental principle. This strategy prioritizes the closeness of essential services and amenities within a short distance to enhance accessibility and minimize the necessity for long commutes.

k. *Esthetics and identity*: The Renaissance and Baroque periods elevated urban design to an expression of political power and cultural identity. Cities such as Florence and Rome were redesigned with grand boulevards, monumental architecture, and expansive public squares that symbolized civic pride and communal life [17]. Principles of symmetry, hierarchy, and axial alignment were employed to create visually striking urban landscapes that reflected the aspirations and values of the ruling elite.

l. *Optimal density and sustainable urbanism*: Sustainable urbanism should aim towards achieving optimal density while considering social sustainability factors to create liveable and vibrant urban environments. This also involve designing compact cities with appropriate transit

m. *Order and organization*: Central to early urban planning principles was the concept of order and organization. Ancient civilizations like those in Mesopotamia and classical Greece developed structured city layouts characterized by gridiron street patterns and centralized public spaces [18]. These principles aimed to create cohesive urban environments that facilitated efficient movement, governance, and commerce, laying the groundwork for systematic urban planning practices.

n. *Functionality and efficiency*: The modernist movements of the early twentieth century introduced principles of functionality and efficiency to urban design. Figures like Le Corbusier advocated for the rationalization of urban spaces through high-rise buildings, separation of functions, and geometric city planning [19]. The modernist vision aimed to address the social and environmental challenges of urbanization, promoting efficient land use, transportation networks, and sanitary conditions but often at the expense of human-scale environments and community cohesion.

o. *Sustainability and resilience*: Contemporary urban design principles increasingly emphasize sustainability and resilience in response to global environmental challenges. Concepts such as sustainable development, smart growth, and biophilic design integrate ecological principles into urban planning practices [23]. These principles advocate for green infrastructure, energy efficiency, and adaptive reuse strategies to enhance urban resilience and mitigate the impacts of climate change. Integrating environmentally friendly practices will help to create sustainable urban environments.

These principles, when applied thoughtfully and holistically, can help create liveable, sustainable, and resilient urban environments that benefit both current and future generations.

## 4. Integrating nature and built environment in design and planning of urban spaces

Incorporating nature into urban design and planning dates back to the nineteenth century and has been advocated by influential figures like Frederick Law Olmsted, Ebenezer Howard, Frank Lloyd Wright, Patrick Geddes, Lewis Mumford, and Ian. These pioneers laid the groundwork for ecological planning, emphasizing the importance of incorporating natural elements into urban spaces [8, 28–33]. This is crucial today for several reasons. It enhances environmental sustainability by reducing urban heat islands, managing stormwater, and improving air quality. Green spaces and natural elements promote well-being, offering residents opportunities for recreation and relaxation. Furthermore, this fosters biodiversity, supporting urban wildlife and creating more resilient ecosystems. The following concepts outline the key strategies for achieving this integration:

### 4.1 Biophilic design

Biophilic design highlights humanity's innate connection to nature by integrating natural elements into the built environment. This approach involves utilizing natural materials, incorporating green spaces, and ensuring ample access to natural light and ventilation [8, 34]. Beyond mere biodiversity, a biophilic city, as described by Beatley, learns from and mimics natural systems, integrates natural forms and imagery into its architecture and urban landscapes, and plans and designs harmoniously with nature. Such cities not only cherish existing natural features but also strive to restore and rehabilitate degraded environments.

Incorporating biophilic design into buildings can be achieved through various strategies that enhance the connection between occupants and nature. Bringing greenery into indoor spaces is a fundamental aspect of biophilic design. This can range from placing potted plants throughout rooms to incorporating larger features such as pergolas adorned with vines or even indoor trees. These natural elements not only improve air quality but also create a calming and esthetically pleasing environment. Employing natural materials in construction and interior decoration fosters a deeper connection to nature. Wood, bamboo, rattan, and cork are excellent choices that bring warmth and texture to a space. These materials can be used for flooring, furniture, wall coverings, and decorative accents, contributing to a more organic and inviting atmosphere. Indoor water features, such as fountains, aquariums, or even small ponds, can significantly enhance the biophilic quality of a building. Research has shown that water elements reduce stress and promote relaxation. In the noisy and hectic environments of urban areas, they also help mitigate noise pollution, creating a tranquil and serene setting. Designing spaces that emulate the natural world can be highly effective. This could involve creating indoor gardens, using green walls, or designing layouts that resemble natural landscapes. Such approaches not only improve the esthetic appeal but also foster a sense of being connected to the outdoors. Maximizing natural light in a building is another key principle of biophilic design. Large windows, skylights, and light wells can be used to bring sunlight into interior spaces. Natural light not only enhances the mood and productivity of occupants but also reduces the need for artificial lighting, promoting energy efficiency. By integrating these elements, buildings can achieve a harmonious balance between the built environment and the natural world, leading to healthier, more enjoyable, and more sustainable living and working spaces. These are described in ways below.

### 4.2 Ecosystem services

Ecosystem services refer to the essential benefits that nature provides, including air and water purification, climate regulation, and recreational opportunities [35]. Urban planning that integrates these services can significantly enhance environmental quality and resilience [36]. By strategically incorporating ecosystem services into urban design, cities can mitigate environmental degradation, support biodiversity, and promote sustainable development [37]. Green infrastructure, such as parks and green roofs, plays a crucial role in improving air quality and managing stormwater effectively [38]. Additionally, integrating natural habitats within urban areas not only provides wildlife habitats but also enhances the esthetic and recreational value of urban landscapes [39]. This comprehensive approach not only addresses current environmental challenges but also prepares cities to adapt to future uncertainties, such as climate change and urbanization [40]. Emphasizing ecosystem services in urban planning underscores the interconnectedness of human well-being and environmental health, fostering resilient and sustainable cities for future generations.

### 4.3 Urban agriculture

Including urban agriculture in city planning supports local food production, bolsters food security, and diminishes the carbon footprint linked with food transportation. Urban farms, rooftop gardens, and community plots additionally serve as green spaces, offer recreational opportunities, and foster urban biodiversity [41].

### 4.4 Wildlife corridors

Creating wildlife corridors in urban areas is crucial for allowing animals to move safely between habitats, thereby promoting biodiversity and ecological resilience. These corridors can take various forms, such as greenways, riverside paths, and interconnected parks, which provide continuous natural pathways for wildlife [42]. Additionally, incorporating native vegetation and minimizing barriers such as roads and buildings can enhance the effectiveness of these corridors, fostering a healthier and more balanced urban ecosystem. By integrating these natural elements into city planning, we not only support wildlife but also create more sustainable and livable urban environments for people.

### 4.5 Climate-resilient planning

Climate-resilient urban planning focuses on mitigating the impacts of climate change by incorporating adaptive strategies into urban design. This approach includes designing buildings and infrastructure that can withstand extreme weather events, such as hurricanes, floods, and heatwaves. It also involves creating buffer zones to protect against flooding and sea-level rise, and enhancing urban green spaces to mitigate the urban heat island effect [43]. Additionally, climate-resilient planning promotes sustainable water management practices, such as the use of permeable surfaces and rainwater harvesting, to reduce runoff and conserve water. By prioritizing these strategies, cities can become more resilient, sustainable, and better equipped to handle the challenges posed by a changing climate.

### 4.6 Community involvement

Engaging local communities in the planning and development of green spaces is essential for ensuring these areas meet the needs and preferences of residents. Involving community members through participatory planning processes not only enhances social cohesion and fosters a sense of ownership but also encourages the sustainable use and maintenance of green spaces [44]. Additionally, involving residents in decision-making helps to identify and address specific local concerns, such as safety, accessibility, and recreational needs. This collaborative approach can lead to more innovative and culturally relevant green spaces that contribute to the overall well-being and quality of life in urban areas. By valuing and incorporating community input, urban planners can create vibrant, inclusive, and enduring green spaces that reflect the unique character and aspirations of the communities they serve.

Integrating nature into urban design and planning is crucial for creating cities that are both environmentally sustainable and supportive of the health and well-being of their residents. By incorporating natural elements, urban planners and designers can develop vibrant, resilient, and liveable urban environments that coexist harmoniously with the natural world. This approach not only enhances the ecological sustainability of cities but also contributes to the overall quality of life, making urban areas more attractive and healthier places to live.

### 4.7 Benefits of integrating nature and the built environment into urban design and planning

There are numerous benefits to incorporating wildlife into urban planning and design that extend beyond environmental sustainability. This holistic approach enhances the quality of life for urban residents, promotes public health, boosts economic vitality, and fosters social cohesion. By weaving natural elements into the fabric of urban environments, cities can become more resilient and adaptable to future challenges. This section delves into the diverse advantages of incorporating nature into urban spaces, illustrating how this integration supports ecological, social, and economic well-being.

### 4.8 Environmental sustainability

To promote environmental sustainability, natural components must be blended into urban areas. In addition to reducing storm water runoff and increasing air quality by filtering pollutants, green areas, urban forests, and green roofs contribute to reduce the urban heat island effect. Sustainable urban drainage systems (SUDS) manage stormwater effectively, reducing the risk of flooding and improving water quality [10, 11, 32].

### 4.9 Public health and well-being

The presence of green spaces and natural environments within urban areas plays a vital role in promoting both physical and mental health. Research indicates that exposure to natural settings can effectively lower stress levels, elevate mood, and enhance overall well-being. Parks and recreational areas also encourage physical activity, which helps combat obesity and related health issues. Moreover, integrating natural elements into urban planning can mitigate noise pollution and enhance air quality, thereby fostering healthier and more pleasant living environments [8, 45, 46]. These findings underscore the importance of incorporating green spaces into urban development to support the holistic health of urban residents.

### 4.10 Biodiversity and ecosystem services

Urban biodiversity thrives through the presence of green spaces and wildlife corridors, which provide essential habitats for a diverse range of plant and animal species. Maintaining this biodiversity is crucial for preserving ecological balance and

*DOI: http://dx.doi.org/10.5772/intechopen.1006805*

enhancing the resilience of urban ecosystems. Moreover, the integration of natural elements into city landscapes strengthens ecosystem services such as pollination, carbon sequestration, and soil stabilization [42]. Incorporating bird and butterfly gardens into urban settings, as highlighted in Ref. [31], offers numerous benefits. These specialized green spaces not only support flower pollination and provide food sources for wildlife but also contribute significantly to environmental conservation efforts. Urban residents reap advantages such as natural pest and weed control, stress reduction from connecting with nature, and increased property values due to the appeal of sustainable environments. Additionally, these gardens serve as effective educational platforms for raising awareness about environmental issues and promoting sustainable practices within urban communities.

### 4.11 Economic benefits

The presence of green infrastructure can boost local economies by increasing property values and attracting businesses and tourists. Esthetic enhancements provided by green spaces make cities more attractive to residents and visitors alike, promoting economic development. In addition to enhancing local food security and reducing transportation costs, urban agriculture initiatives play a crucial role. Furthermore, green infrastructure helps in minimizing expenses related to stormwater management and energy consumption [41, 47].

### 4.12 Social cohesion and community engagement

Green spaces and communal gardens serve as social hubs where residents can gather, interact, and engage in community activities. These spaces foster a sense of community and belonging, enhancing social cohesion and reducing social isolation. Participatory planning processes that involve residents in the design and maintenance of green spaces empower communities and encourage sustainable urban living [44].

### 4.13 Resilience and climate adaptation

Integrating nature into urban design offers significant benefits in enhancing urban resilience to climate change and environmental challenges. By incorporating green infrastructure such as parks, green roofs, and urban forests, cities can better adapt to extreme weather events. For example, natural elements such as trees and green spaces provide cooling effects during heatwaves, while features such as rain gardens and permeable surfaces help manage stormwater and mitigate flooding caused by heavy rainfall [43]. This holistic approach not only addresses immediate climate risks but also fosters long-term sustainability and enhances the overall livability of urban areas, creating environments more capable of enduring and recovering from environmental disruptions.

### 4.14 Esthetic and cultural value

Natural elements enhance the esthetic quality of urban environments, creating visually appealing and culturally significant spaces. Parks, gardens, and green roofs contribute to the beauty and character of cities, reflecting cultural heritage and providing spaces for cultural expression and activities [8].

Integrating nature into urban design and planning yields comprehensive benefits that touch upon every aspect of urban life. By embracing these principles, cities can foster environments that are not only sustainable and resilient but also vibrant and conducive to the well-being of their residents.

## 4.15 Challenges in integrating nature

Nature-inspired urban architecture and planning provide numerous benefits, but a number of difficulties must be addressed before it can be fully realized. As highlighted in the previous section, blending natural systems with urban infrastructure promotes sustainability, enhances public health, supports biodiversity, boosts economic vitality, fosters social cohesion, increases resilience, and adds esthetic and cultural value. However, achieving these benefits requires navigating various obstacles, including regulatory, financial, and technical barriers. This section investigates the challenges of bringing nature into urban environments in order to develop more sustainable, resilient, and habitable cities. It accomplishes this by emphasizing key challenges and proposing alternative remedies.

### *4.15.1 Regulatory and policy barriers*

Implementing green infrastructure often requires navigating complex regulatory environments and policy frameworks. Existing zoning laws and building codes may not accommodate innovative designs that incorporate natural elements [11]. Policymakers and planners need to collaborate to update regulations and create supportive policies that encourage sustainable urban design [8].

### *4.15.2 Financial constraints*

The initial costs of integrating green infrastructure can be high, posing a significant challenge for municipalities and developers. Securing funding for these projects often requires demonstrating long-term economic benefits and cost savings through mechanisms such as public-private partnerships, grants, and incentives for sustainable development [42]. Financial constraints are frequently not related to environmental finance, but rather the outcome of a low-income base or insufficient tax collection. The financial capability of cities is determined by the national constitutional and legislative framework, which influences both expenditure duties and revenue sources.

### *4.15.3 Technical and logistical challenges*

Designing and implementing green infrastructure involves overcoming technical challenges related to urban space constraints, soil quality, water management, and plant selection [32]. Urban planners and designers must work with experts in ecology, landscape architecture, and engineering to develop feasible and effective solutions [45].

### *4.15.4 Maintenance and management*

Green infrastructure requires ongoing maintenance to ensure its functionality and esthetic appeal. Developing sustainable maintenance plans and engaging community members in the stewardship of green spaces can help address this challenge. Public education and community involvement are crucial for the success of these initiatives [44].

#### *4.15.5 Equity and access*

Ensuring equitable access to green spaces and the benefits of nature-integrated urban design is a significant concern. Planners must prioritize inclusive strategies that address disparities in access to green infrastructure across different neighborhoods and communities, particularly in underserved areas [41].

#### *4.15.6 Climate change and urbanization pressures*

The effects of climate change and the massive urbanization that is taking place make them more difficult to integrate nature into cities. Planners must develop adaptive strategies that can withstand extreme weather events and changing environmental conditions while accommodating population growth and urban expansion [43]. Despite these challenges, there are numerous opportunities for innovation and progress. Advances in technology, increased awareness of environmental issues, and growing public support for sustainable practices provide a fertile ground for integrating nature into urban design and planning. Collaborative efforts among government agencies, private sector stakeholders, and communities can lead to the development of innovative solutions that balance urban growth with ecological sustainability. As a result, while incorporating nature into urban settings has its difficulties, it also has tremendous potential for developing cities that are lively, inclusive, resilient, and sustainable. Addressing these challenges requires a multifaceted approach, leveraging policy changes, financial investments, technical innovations, and community engagement. The following sections will delve overcoming the challenges.

### 4.16 Overcoming barriers to nature integration

Integrating nature into urban design and planning is essential for creating sustainable and liveable cities, but it is not without its challenges. Overcoming these barriers requires strategic approaches and collaborative efforts from various stakeholders. This section outlines several key strategies for addressing regulatory, financial, technical, maintenance, equity, and climate-related obstacles to integrating nature into urban environments.

#### *4.16.1 Regulatory and policy reforms*

Updating and revising existing regulations and policies is crucial for facilitating nature integration in urban areas. Planners and policymakers must work together to create a supportive legal framework that encourages sustainable design practices. This includes revising zoning laws, building codes, and land-use policies to allow for green infrastructure, such as green roofs, urban agriculture, and wildlife corridors [11]. Moreover, policies that incentivize green building practices, such as tax breaks or grants for developers who incorporate natural elements, can further promote ecological urban design [8].

#### *4.16.2 Innovative financing mechanisms*

Addressing financial constraints involves exploring various funding sources and financing models. Public-private partnerships can be instrumental in pooling resources and sharing risks associated with green infrastructure projects.

Additionally, implementing green bonds and environmental impact bonds can attract investment by linking financial returns to environmental performance outcomes [39]. Grants from government agencies and non-profit organizations dedicated to sustainability can also provide essential funding for nature integration projects.

#### *4.16.3 Technical solutions and expertise*

Overcoming technical challenges requires leveraging expertise from multiple disciplines, including ecology, landscape architecture, and civil engineering. Collaborative design processes that involve experts in these fields can lead to innovative solutions for integrating nature into urban areas. For instance, employing advanced modeling and simulation tools can help design effective stormwater management systems and optimize the placement of green spaces to maximize environmental benefits [32]. Adopting best practices from globally successful initiatives can also offer insightful information and direct local implementations.

#### *4.16.4 Sustainable maintenance and management*

Developing sustainable maintenance plans is vital for the long-term success of green infrastructure. This includes engaging community members and local organizations in the stewardship of green spaces. Residents that participate in the community feel more accountable and own their land, which reduces maintenance costs [44]. Community members can attend training sessions and seminars to learn how to properly maintain their green infrastructure.

#### *4.16.5 Promoting equity and access*

Ensuring that all residents have access to the benefits of integrated nature requires intentional and inclusive planning. Planners must prioritize green infrastructure projects in underserved neighborhoods and work to eliminate disparities in access to green spaces. Community engagement and participatory planning processes are essential for understanding the needs and preferences of diverse populations and ensuring that green infrastructure projects benefit all residents equitably [41].

#### *4.16.6 Adapting to climate change and urbanization*

Climate change and rapid urbanization present ongoing challenges that necessitate adaptive and resilient planning strategies. Urban planners must design green infrastructure that can withstand extreme weather events and accommodate future population growth. To mitigate the effects of climate change, incorporate climate resilience into the design process and use natural solutions such as wetlands and green corridors [43]. In the face of changing environmental conditions, the ability to adapt and be flexible is critical for making cities habitable and sustainable.

#### *4.16.7 Harnessing technology and innovation*

Advances in technology offer new opportunities for integrating nature into urban spaces. For instance, smart city technologies can optimize the management of green infrastructure, using sensors and data analytics to monitor environmental conditions and improve maintenance efficiency [45]. Additionally, innovative design approaches, such as biomimicry, can inspire solutions that emulate natural processes and enhance the functionality of urban ecosystems.

Overcoming the barriers to integrating nature into urban design and planning requires a multifaceted approach explained above. By addressing these challenges collaboratively, cities can create vibrant, resilient, and sustainable urban environments that harmoniously blend natural and built systems.

## 5. Global examples of successful nature and built environment

It is critical to perform a complete examination of diverse global situations by examining how nature and the built environment have been incorporated into urban design and planning. We can learn a lot about creative concepts and global best practices by examining them. To improve our understanding and implementation of nature integration in cities, this research focuses on the prospect of increasing the number and variety of these situations.

The famed Barbican Centre in London, a pioneering example of biophilic design, was constructed in the 1980s. The Barbican, a well-known example of brutalist architecture, provides an amazingly harmonious urban environment by mixing angular concrete forms with a variety of natural materials (**Figure 1**). Lakes, both man-made and natural, provide a more soft contrast to the Barbican's normally austere form while also serving as a habitat for a variety of wildlife. The Barbican Conservatory, which houses over 1500 different plant and tree species, is one of the building's most prominent attractions. This lush, green space provides a striking contrast to the

**Figure 1.**
*The Barbican Centre. Source: Image from UCEM Article: https://www.ucem.ac.uk/whats-happening/articles/biophilia-examples-built-environment/.*

surrounding concrete structures, illustrating the powerful impact of integrating natural elements into urban design. The conservatory, along with the surrounding gardens and water features, enhances the esthetic appeal of the estate while also promoting biodiversity and environmental sustainability. The Barbican Centre exemplifies how biophilic design can transform urban spaces, making them more livable and enjoyable for residents and visitors. By incorporating extensive green spaces and water elements into its design, the Barbican offers a serene and refreshing environment that stands in stark contrast to the typical urban landscape. This novel approach to urban design emphasizes the long-term worth and benefits of incorporating nature into the built environment, laying the groundwork for future developments that prioritize ecological harmony and human well-being.

Another notable example is the Brackweder Gymnasium in Bielefeld in Germany: This serves as a prominent example of the successful integration of nature with the built environment. This historical building, nestled within a green landscape, exemplifies how architectural design can harmoniously coexist with natural elements.

Aside from boosting the building's esthetics, planting vegetation, trees, and bushes in the surrounding region promotes biodiversity and environmental sustainability. This approach to urban design provides numerous benefits, including improved air quality, natural temperature regulation, and a serene environment for students and staff. The photograph of the Brackweder Gymnasium, taken on 25 June 2024, captures this beautiful synergy between nature and architecture (**Figure 2**). This integration reflects a growing trend in urban planning that emphasizes the importance of creating eco-friendly, liveable spaces within urban settings.

The image presented in **Figure 3** showcases an indoor space with natural elements, illustrating the integration of nature into the built environment. The room features potted plants and exposed wooden beams, which together foster a deeper connection to nature. It is an esthetically pleasing mix that also adds to a cozier, friendlier atmosphere. Integrating natural materials and greenery within indoor spaces is a key aspect of biophilic design. Plants improve air quality and create a visually pleasing environment, while natural materials like wood bring warmth and texture to the space. The image exemplifies how incorporating these elements can transform an interior setting, promoting well-being and a sense of tranquility. To increase tenants'

**Figure 2.**
*Brackweder Gymnasium, Bielefeld, Germany. Source: Photo by the author.*

**Figure 3.**
*A room in a residential apartment. Source: Photo by the author.*

quality of life, biophilic design concepts stress the importance of connecting indoor and outdoor habitats. Designers may create places that are both utilitarian and compassionate by including natural light, water elements, and natural materials. This approach is increasingly recognized for its benefits, which include reduced stress levels, enhanced mood, and overall better mental and physical health.

## 6. Conclusion: the future of integrating nature and built environment in urban design and planning

In conclusion, this chapter underscores the transformative impact of integrating natural components with urban infrastructure. Throughout this research, we have looked at the underlying concepts, tactics, difficulties, and future directions that constitute this dynamic field. Integrating nature into urban design has numerous benefits. As Li shows, incorporating green infrastructure can improve environmental sustainability by encouraging biodiversity and reducing climate impacts, especially in cities like Los Angeles [48]. However, Francis and Lorimer stress that conserving biodiversity in urban environments presents considerable problems, raising concerns about the long-term viability of such efforts [49]. According to Ref. [47], green spaces help to reduce stress, encourage physical exercise, and improve air quality. However, Rigolon claims that not all populations can gain equally from these benefits, citing inequities that may threaten overall public health outcomes [50]. Green infrastructure has been established in research to promote social equality and inclusivity by ensuring that all residents have access to public spaces [50–52]. This should be accomplished by ensuring that all inhabitants have access to parks, open spaces, and cultural landscapes, as well as cultivating resilient neighborhoods capable of responding to the difficulties brought by urbanization and climate change. However, another study contradicts this, showing out that green infrastructure can occasionally lead to gentrification, which exacerbates rather than lessens social inequality [53].

This demonstrates that accomplishing diverse goals necessitates overcoming several barriers. Navigating the problems of legislative limits, financial barriers, and community engagement is critical, as noted by Li [48], who emphasizes the necessity for new ways and collaborative partnerships to overcome these barriers. A research by Wolch et al. [47] also advocates for incorporating smart technologies into green infrastructure to improve sustainability, but warns that such developments have the potential to worsen inequality if not carefully managed. Ultimately, integrating nature into urban design is not merely a trend but a necessary paradigm shift towards creating cities that are environmentally harmonious and socially inclusive. By considering the insights and critiques offered by these studies, urban planners can better navigate the complexities of this integration, ensuring that cities evolve into vibrant, livable spaces that reflect our collective responsibility to nurture both the built and natural environments.

## Acknowledgements

This contribution was written during my stay at Centre for Interdisciplinary Research (ZiF), Bielefeld University, Bielefeld, Germany, as a Norbert Elias Fellow on the research Internalizing borders.

The author acknowledges the use of ChatGPT during the initial brainstorming of the article.

## Conflict of interest

The author declares no conflict of interest.

## Author details

Deborah Bunmi Ojo[1,2]

1 Department of Urban and Regional Planning, Obafemi Awolowo University, Ile-Ife, Nigeria

2 Center for Interdisciplinary Research (ZiF), Bielefeld University, Bielefeld, Germany

*Address all correspondence to: dbojo@oauife.edu.ng

## References

[1] de Vries B, Vincent T, Henri AA. Interactive urban design using integrated planning requirements control. Automation in Construction. 2005;**14**(2):207-213. ISSN 0926-5805. DOI: 10.1016/j.autcon.2004.07.006

[2] Gehl J. Cities for People. Washington: Island Press; 2010

[3] Loukaitou-Sideris A, Banerjee T. In: Moskowitz M, editor. Urban Design Downtown: Poetics and Politics of Form. Berkeley: University of California Press; 1998. Vol. 36(2). p. 380. ISBN 052020930

[4] Hall P. Cities of Tomorrow: An Intellectual History of Urban Planning and Design in the Twentieth Century. Hoboken, New Jersey: Wiley-Blackwell; 2002

[5] Roberts M, Eldridge A. Planning the Night-Time Economy. London: Routledge; 2009

[6] United Nations. 68% of the world population projected to live in urban areas by 2050, says UN. 2018. Available from: https://www.un.org/development/desa/en/news/population/2018-revision-of-world-urbanization-prospects.html

[7] Sirle S, Zenia K. Spatial interventions and built environment features in developing age-friendly communities from the perspective of urban planning and design. Cities. 2023;**141**:104417. ISSN 0264-2751. DOI: 10.1016/j.cities.2023.104417

[8] Beatley T. Biophilic Cities: Integrating Nature into Urban Design and Planning. Washington, DC: Island Press; 2011

[9] Chiesura A. The role of urban parks for the Sustainable City. Landscape and Urban Planning. 2004;**68**(1):129-138

[10] Gill SE, Handley JF, Ennos AR, Pauleit S. Adapting cities for climate change: The role of the green infrastructure. Built Environment. 2007;**33**(1):115-133

[11] Newman P, Beatley T, Boyer H. Resilient Cities: Overcoming Fossil Fuel Dependence. Washington, DC: Island Press; 2017

[12] Bryant C. Book review: Spatial revolution: Architecture and planning in the early Soviet Union. Journal of Planning History. 2024;**23**(2):154-156. DOI: 10.1177/15385132231194952

[13] Norgaard S. From "citizen Jane" to an institutional history of power and social change: Problematizing urban Planning's Jane Jacobs historiography. Journal of Planning History. 2023;**22**(2):95-118. DOI: 10.1177/15385132211070512

[14] Shevah D, Kallus R. Past forward: Planning in the light of historical knowledge. Journal of Planning History. 2016;**15**(1):29-45. DOI: 10.1177/1538513214553909

[15] Ayyoob S. From Garden City to eco-urbanism. The quest for sustainable neighborhood development. Sustainable Cities and Society. 2016;**20**:1-16. ISSN 2210-6707. DOI: 10.1016/j.scs.2015.09.002

[16] Hirt SA. Premodern, modern, postmodern? Placing new urbanism into a historical perspective. Journal of Planning History. 2009;**8**(3):248-273. DOI: 10.1177/1538513209338902

[17] Raven J, Stone B, Mills G, Towers J, Katzschner L, Leone M, et al. Urban planning and design. In: Rosenzweig C, Solecki W, Romero-Lankao P, Mehrotra S, Dhakal S, Ibrahim SA, editors. Climate Change and Cities: Second Assessment

Report of the Urban Climate Change Research Network. New York: Cambridge University Press; 2018. p. 139

[18] Mumford L. The City in History: Its Origins, its Transformations, and its Prospects. New York: Harcourt Brace & World Inc; 1961

[19] Fishman R. In: Howard E, Wright FL, Corbusier L, editors. Urban Utopias in the Twentieth Century. Cambridge, Massachusetts, London, England: MIT Press; 1982

[20] Jacobs J. The Death and Life of Great American Cities. New York City, United State: Random House; 1961

[21] Gordon DLA, Virginillo M. The post-war revival of Canadian planning: Assessing the impact of the community planning Association of Canada. Journal of Planning History. 2024;**23**(2):110-125. DOI: 10.1177/15385132231222853

[22] Paine G, Thompson S, Prior J, Connon I, Kent JL. Bringing history forward: Learning from historical context when translating contemporary health evidence into planning practice. Journal of Planning History. 2022;**21**(2):107-131. DOI: 10.1177/1538513220977456

[23] Ahern J. From fail-safe to safe-to-fail: Sustainability and resilience in the new urban world. Landscape and Urban Planning. 2011;**100**(4):341-343

[24] Smith ME. Form and meaning in the earliest cities: A new approach to ancient urban planning. Journal of Planning History. 2007;**6**(1):3-47. DOI: 10.1177/1538513206293713

[25] Hashem D, Afshin S. Principles of just urban land use planning. Land Use Policy. 2024;**141**:107132. ISSN 0264-8377. DOI: 10.1016/j.landusepol.2024.107132

[26] Amir R, Ayyoob S, Ali S. The 15-minute city: Urban planning and design efforts toward creating sustainable neighborhoods. Cities. The International Journal of Urban Policy and Planning. 2023;**132**:104101. ISSN 0264-2751. DOI: 10.1016/j.cities.2022.104101

[27] Moreno C, Zaheer A, Didier C, Catherine G, Florent P. Introducing the "15-Minute City": Sustainability, resilience and place identity in future post-pandemic cities. Smart Cities. 2021;**4**(1):93-111. DOI: 10.3390/smartcities4010006

[28] Dunnett N, Kingsbury N. Planting Green Roofs and Living Walls. Portland, Oregon: Timber Press; 2008

[29] Well F, Ludwig F. Development of an integrated design strategy for blue- green architecture. Sustainability. 2021;**13**:7944. DOI: 10.3390/su13147944

[30] Russo A, Cirella GT. Urban sustainability: Integrating ecology in City design and planning. In: Sustainable Human–Nature Relations: Environmental Scholarship, Economic Evaluation, Urban Strategies. Advances in 21st Century Human Settlements. Singapore. ISBN: 9789811530494: Springer Singapore; 2020

[31] Firdevs G. Integrating ecosytem landscapes in cityscape: Birds and butterflies. In: Ozyavuz M, editor. Advances in Landscape Architecture. London, United Kingdom: InTechOpen; 2013. p. 940. DOI: 10.5772/55753

[32] Bolund P, Hunhammar S. Ecosystem Services in Urban Areas. Ecological Economics. 1999;**29**(2):293-301

[33] Thompson GF, Steiner F. Ecological Design and Planning. New York: John Wiley & Sons; 1997

[34] Kellert SR. Dimensions, elements, and attributes of biophilic design. In: Kellert SR, Heerwagen JH, Mador ML, editors. Biophilic Design: The Theory, Science, and Practice of Bringing Buildings to Life. New York: Wiley; 2008. pp. 3-19

[35] Millennium Ecosystem Assessment (MEA). Ecosystems and Human Well-Being: Synthesis. Washington, DC: Island Press; 2005

[36] Elmqvist T et al. Urbanization, Biodiversity and Ecosystem Services: Challenges and Opportunities. Netherlands: Springer Science & Business Media; 2013

[37] Tzoulas K et al. Promoting ecosystem and human health in urban areas using green infrastructure: A literature review. Landscape and Urban Planning. 2007;**81**(3):167-178

[38] Ahern J. Urban landscape sustainability and resilience: The promise and challenges of integrating ecology with urban planning and design. Landscape Ecology. 2013;**28**(6):1203-1212

[39] Benedict MA, McMahon ET. Green Infrastructure: Linking Landscapes and Communities. Washington, DC: Island Press; 2012

[40] Grimm NB et al. Global change and the ecology of cities. Science. 2008;**319**(5864):756-760

[41] Mougeot LJA. Growing Better Cities: Urban Agriculture for Sustainable Development. Canada: IDRC; 2006

[42] Benedict MA, McMahon ET. Green Infrastructure: Linking Landscapes and Communities. Washington, DC: Island Press; 2006

[43] IIPCC. Climate Change 2014: Impacts, Adaptation, and Vulnerability. Cambridge, United Kingdom: Cambridge University Press; 2014

[44] Arnstein SR. A ladder of citizen participation. Journal of the American Institute of Planners. 1969;**35**(4):216-224

[45] Kellert SR. Biophilic Design: The Theory, Science, and Practice of Bringing Buildings to Life. New York: John Wiley & Sons; 2015

[46] Maas J, Verheij RA, Groenewegen PP, De Vries S, Spreeuwenberg P. Green space, urbanity, and health: How strong is the relation? Journal of Epidemiology & Community Health. 2006;**60**(7):587-592

[47] Wolch JR, Byrne J, Newell JP. Urban green space, public health, and environmental justice: The challenge of making cities 'just green enough'. Landscape and Urban Planning. 2014;**125**:234-244

[48] Li J. The Interplay of Urbanization, Climate Change, Green Infrastructure, and Biodiversity Preservation in Los Angeles. New York: Columbia University; 2024

[49] Francis RA, Lorimer J. Urban reconciliation ecology: The potential of living roofs and walls. Journal of Environmental Management. 2011;**92**(6):1429-1437

[50] Rigolon A. A complex landscape of inequity in access to urban parks: A literature review. Landscape and Urban Planning. 2016;**153**:160-169

[51] Anguelovsk I. Justice in urban greening and green gentrifcation. In: Villamayor-Tomas S, Muradian R, editors. The Barcelona School of Ecological